An Introduction to the Regulation of Primary Metabolic Pathways

An Introduction to the Regulation of Primary Metabolic Pathways

Adhikarla S. Rao (A.S. Rao)

KALPAZ

First Published, 2017

Published by

Kalpaz Publications
C-30, Satyawati Nagar,
Delhi – 110052
E-mail: kalpaz@hotmail.com
Ph.: 9212142040

Printed at: G. Print Process, Delhi

Cataloging in Publication Data—DK
Courtesy: D.K. Agencies (P) Ltd. <docinfo@dkagencies.com>

Rao, Adhikarla S., author.
An introduction to the regulation of primary metabolic pathways / Adhikarla S. Rao.
pages cm
Includes bibliographical references.
ISBN 9789386397102

1. Metabolism—Regulation. I. Title.

QH521.R36 2017 DDC 572.4 23

Contents

PART - III

Regulation of Lipid Metabolism

PART - IV

Regulation of Amino acid Metabolism

PART - VII

Zymogen activation, Protein targeting and Protein turnover

PART - VIII

A Brief Introduction to Gene Regulation in Prokaryotes and Eukaryotes and Signal Transduction in Mammals and Plants

To All My

Teachers, Colleagues and Students

Acknowledgements

At the outset I wish to wholeheartedly acknowledge Dr S. Sivaramakrishnan, Head of the Department of Biotechnology, Bharathidasan University and Honourable Syndicate Member of our university representing the Chancellor of the University (Governor of Tamil Nadu state) who had been a great source of help, security and inspiration for me. He is the spirit behind writing this book. I express my profuse thanks to him.

Words are not sufficient to acknowledge all my teachers, colleagues and students in different departments of different institutes in different states of the country without whom I would not have learnt anything worthwhile. Hence I am wholeheartedly dedicating this book to all of them.

My interactions at different institutes at Switzerland, UK, France, Germany, Netherlands and Singapore have enlightened me a lot both about pedagogy, research and policy matters which have been useful to me in writing this book, apart from rendering my prescribed primary duties.

I wish to express my sincere gratitude to all my colleagues in the Department of Biotechnology and sister departments of School of Life Sciences and other Schools. The administration- both academicians and secretarial- of the university at all levels, has been extremely kind to me.

Dr A. S. Rao (Adhikarla S. Rao).

Professor (Retired),

Former:

Head Department of Biotechnology,
Chair - School of Life sciences,
In-charge- **Coordinator**, 5Yr Integrated Life Science Program,
Syndicate Member Bharathidasan University,
Honorary Professor, Department of Biotechnology,
Bharathidasan University, Tiruchirappalli- 620024,Tamil Nadu, India
Mob: +919789272567; email: adhikarla@gmail.com/asraobio@gmail.com

PREFACE

Students who are in undergraduate programs either 3-year or 4-year programs or studying in 5-year integrated Post graduate programs in different disciplines, find it very difficult to get an overview/basics of metabolic regulation and in fact find it daunting to know all the details going through bulky text books. There are also students pursuing UG and PG programs in Engineering and Technology discipline related to Biotechnology. These students are not in need of exhaustive details but only the core details. In addition to students studying in regular mode in India there are many UG and PG programs in life sciences offered in a distance education mode also called correspondence courses. India has many distance education centres both in India and abroad which is likely to be further expanded. These students are badly in need of an overview of the subject/topics. For the above mentioned this smaller text can serve as a good reference book. So the text is simplified by not mentioning the structures of biomolecules and complete metabolic pathways but only the regulatory steps that too simplified without structures. Details mentioned in the text will serve as a good pre-requisite and background for additional and advanced studies. Further students appearing for oral interviews for advanced studies both within the country and by telephone/skype from foreign countries need the core concepts rather than exhaustive details. In particular this text would be highly useful for such students.

In view of the above, the purpose of this text book is not to give encyclopedic details but give the salient points. Hence the contents are described accordingly, so that students can get easily familiarised with an overview of metabolic regulation. The primary emphasis is on the regulation of metabolic pathways through the regulation of enzyme activity—also termed as short-term regulation or fine control of enzyme activity (i.e. seconds to minutes). As metabolic pathways can also be regulated by enzyme amounts—termed as long-term control or coarse control (i.e. hours to days)—a brief overview as to how regulation of enzyme amounts may occur has also been described (Genetic Regulation in both Prokaryotes and Eukaryotes and Signal Transduction both in mammalian and plant systems). These details students usually learn in a molecular biology and/or genetics courses rather than in a biochemistry course. However, the brief but core discussion made here will help the students to study the genetic regulation or gene expression in their molecular biology/genetics courses. The details in the book give an idea as to how enzyme levels/amounts can be regulated and can influence a metabolic pathway.

Details mentioned in the book will help the student to study metabolic regulation at an advanced level in their academic career and to understand the biochemical unity in metabolic regulation among various organisms.

There are 6728 enzymes as on 2016 (http://www.enzyme-database.org/stats.php) and we don't discuss in our curriculum nor any text book—anywhere in the world—discusses the regulation of all these enzymes. So it is certain that the regulation of so many enzymes is yet to be investigated. This constitutes an important and vast area of research in future. Further in addition to enzymes, there are thousands of non-enzymatic proteins whose synthesis, activity and role needs to be investigated. Here - like in other text books - discussed the regulation of only few enzymes from the above mentioned number.

Perfect regulation of metabolism is necessary for normal growth, maintenance and development of any organism. Further living organisms have to regulate their metabolic pathway responding to both internal or external needs and pressures. Abnormality in the regulation of metabolic pathways may result in disorders/deficiencies/diseases/decreased productivity in higher organisms and can affect the performance of the organism. By understanding of the above abnormal conditions, therapeutic approaches or remedial measures can be thought of and required productivity increased in organisms of interest.

An understanding of the metabolic regulation helps in understanding the interlinking of different metabolic pathways and help in designing strategies to redirect metabolism to achieve desired biotechnological goals- that is to say redirecting the metabolism towards the synthesis of specific desired products or stopping or reducing the synthesis of undesirable metabolic products.

The details of metabolic regulation we have is from a tiny number of the total organisms and species existing on earth which are in multi-millions. Hence study of other organisms and species may reveal some novel features of metabolic regulation in addition to the existing known features. Hence a huge number of organisms need to be investigated in future. This constitutes another important and vast area of research in future. In this book available salient examples have discussed from the study of only a few groups of organisms as available in the standard text books and literature.

Even from the existing studies it is evident that regulation of metabolic pathways may vary from organism to organism—say human, plant and microorganism. Again among the multitude of plants and microorganisms there can be variations. Further the regulation can vary within the same organism in different parts.

Humans suffer from a huge number of congenital abnormalities or inborn errors of metabolism and pathological states. Hence in them the regulation of metabolic pathways can be different. However, some basic modes of regulation are common to all organisms and within an organism.

In the present book regulation- that too at the enzyme activity level and an overview at the enzyme amounts level - as it works under normal conditions and internal needs is primarily described—but not in response to different external pressures but for a casual reference in some topics.

I sincerely hope that the book would be highly useful to a large number of students from various backgrounds. Feedback from all is most welcome, for which I would be highly grateful.

Dr A. S. Rao (Adhikarla S. Rao)

adhikarla@gmail.com; asraobio@gmail.com

FOREWORD

It gives me great pleasure and satisfaction to express myself about my esteemed colleague, Dr A. S. Rao, Professor (Retired) in my Department.

In view of my teaching experience and feedback from students in various programs and institutes, it is I who motivated Professor A. S. Rao to take up this venture. I have read the book critically and completely satisfied with the contents.

Relatively recently, there had been a change in the educational policy. Now, specialised Undergraduate Programs in Biotechnology, Microbiology and Biochemistry are being offered apart from the traditional programs specialising in Botany and Zoology, in a huge number of institutes. The students strength in the above mentioned specialised programs is very large. The book entitled - An Introduction to Regulation of Primary Metabolic Pathways - as I expected would serve the needs of Undergraduate students specialising in the above mentioned subjects and also in other disciplines of life sciences as information related to all systems plants, mammals and microorganisms have been discussed. Addition of information on genetic regulation and some applications and on an important enzyme - Phenylalanine ammonia lyase - involved in secondary metabolism, considerably enhances the utility of the book. In view of the above the book also serves the need for the Master's students also in the initial stages in various programs.

The book gives the core concepts of metabolic regulation on the basis of which students can appreciate how a cell or organism sustains itself at molecular level. The contents also would help the students to understand advanced courses and build up further and know how to exploit metabolic pathways towards desired biotechnological goals. The concepts have been explained without the students facing the problem of giving attention to the structures of bio-molecules or every reaction that goes on in a cell. Such information students can learn from other courses in the program depending on their requirement and this book serves as a useful reference for the same and more importantly to revise the core concepts sieving from a large body of information on metabolic pathways and their regulation.

In India we have a large system of distance education in which life science programs are offered and the book also serves as an easy and good background material for them. The book has been written in a manner where students completing their schooling in vernacular medium (Non-English medium) and joining undergraduate programs in English medium can also understand without much difficulty.

I strongly recommend the book for students of all branches of Life Sciences both at Undergraduate and Post Graduate level.

Dr S. Sivaramakrishnan
Head, Department of Biotechnology,
Bharathidasan University,
Tiruchirappalli-620024

GLOSSARY

A

ABA (Abscisic acid)-C15 sesquiterpene and one of the five major plant hormones. Plays a role in many developmental processes.

Abiotic stresses—refers to stresses like drought, salinity etc; stresses not caused by any biological organism.

ACC —1-aminocyclopropane-1-carboxylate, the immediate precursor for ethylene formation catalysed by the enzyme ACC oxidase.

ACC synthase–the enzyme that synthesises ACC from SAM(S-adenosyl-methionine)

Adenyl cyclase—the enzyme that synthesises 3′, 5′ -cyclic AMP from ATP.

Adenylate energy charge—a measure of energy status of the cell which is numerically equivalent to mole fraction of ATP plus half the mole fraction of ADP.

Adenylylation—the addition of 5′AMP group to an enzyme/protein on the side chain of tyrosine amino acid. One of the post-translational covalent modifications involved in the regulation of enzyme activity. Donor for AMP is ATP and the other product is inorganic pyrophosphate (PPi). This reaction is catalysed by adenylyl transferase.

ADP ribosylation—the addition of ADP- ribose to an enzyme/protein, usually on the amino acid residues of lysine, arginine, glutamine and cysteine and diphthamide (a modified histidine residue). One of the post-translational covalent modifications involved in the regulation of enzyme/protein activity. Donor for ADP-ribose is NAD^{+}.

Alkaloids—basic, heterocyclic nitrogen containing plant secondary compounds synthesised from different amino acids. They are pharmacologically active and also act as defensive metabolites for plants against biotic stresses.

Allopurinol—an inhibitor of the enzyme xanthine oxidase in purine catabolism leading to the formation and accumulation of uric acid responsible for the disease of joints, called Gout. It structure is similar to the purine hyoxanthine.

Allosteric regulation—a very important mode of regulation of enzyme activity in metabolic pathways. It involves non-covalent binding of small metabolites which bind at a site other than the substrate binding site of the enzyme. Binding of the metabolite causes a change in the conformation of the enzyme and may either increase or decrease the enzyme activity. Accordingly, they are called allosteric activators and inhibitors respectively.

Alternative splicing–the term refers to the more than one way of splicing i.e cutting and joining of an mRNA molecule giving rise to different matured mRNA molecules which may be translated to different proteins.

Amphibolic pathway—the term refers to a pathway which serves both as a catabolic pathway and also as an anabolic pathway. The intermediates of a catabolic pathway can be used as precursors for the synthesis of other metabolites. For example, both glycolysis and TCA cycle which students often refer to as catabolicpathways, indeed are amphibolic pathways as some of their intermediates can be used for the synthesis of other metabolites.

Amplification of the signal response–an important step in the signal transduction cascade/process. The initially perceived signal is amplified further to bring out the final response effectively.

Amyloplasts— colourless plastids in plants storing starch.

Anaplerotic reactions—also called replenishing or refilling reactions. If any intermediate from a pathway is drained off for other needs, then the pathway may not proceed to completion. In such cases, some special reactions supply the drained off metabolite. Such reactions are called anaplerotic reactions.

Antisense RNA—an RNA molecule that is complementary to a segment of another mRNA whose activity it inhibits. The phenomenon is termed antisense inhibition or suppression. Antisense RNA's are naturally occurring in the organism and can also be chemically synthesised.

Arachidonic acid—it is a 20-carbon polyunsaturated fatty acid, specifically referred to as 5,8,11,14 eicosatetraenoicacid.

A precursor for several pharmacologically and physiologically important compounds, namely prostaglandins, thromboxane's etc.

Arginase—the terminal reaction in the formation of urea in the urea cycle involves the hydrolysis of arginine into ornithine and urea catalysed by the enzyme arginase.

Argonuate—a family of proteins which bind to different classes of small non-coding RNA's and guide them to their specific targets. Hence they play a catalytic role in the RNA induced silencing complex(RISC).

Aromatic amino acid family—phenylalanine, tyrosine and tryptophan.

Aspartate amino acid family—aspartate, asparagine, methionine, lysine andthreonine.

Attenuation—one of the modes of regulation of bacterial operons. Example-Tryptophan operon. Sequence at which termination of transcription occurs is called attenuator which occurs before the start of the first structural gene. Whether transcription is to be continued or not is decided by differential of formation of stem loop structures of the already transcribed mRNA.

Auto catalytic cycle—a cyclic pathway in which the starting metabolite is regenerated at the end of the cycle to participate in another round of cycle. Examples-regeneration of RuBp (Ribulose1,5, bis phosphate in Calvin cycle and oxaloacetate in TCA cycle.

Auxins—one of the five major plant hormones. Indole derivatives and synthesised from tryptophan.

Play a role in many developmental processes. Exploited in tissue culture technology.

B

Beta-oxidation of fatty acids—major energy yielding pathway from stored lipids. The beta carbon of the fatty acid is oxidised in the process with acetyl-CoA as the end product. Occurs in mitochondria, glyoxysomes/peroxisomes.

Bifunctional polypeptide—a polypeptide with two different enzymatic activities i.e both enzymatic activities reside on the same polypeptide.

For example-PFK2 (phosphofructokinase2) and F2,6, BPase (fructose 2,6 bisphosphatase). Many other examples are there.

Biological nitrogen fixation—the conversion of atmospheric nitrogen by certain microorganisms in the root nodules of legume plants. Nodules may occur on stems also. Free living microorganisms also can fix atmospheric nitrogen.

Biotic stresses—stresses caused by biological agents like insects, bacteria, fungi etc.

Both gluconeogenic and ketogenic amino acids—it refers to those amino acids which can give rise to both glucose and also ketone bodies.

Branched amino acids—Isoleucine, leucine and valine, whose side chains are branched.

Branching enzyme—the enzyme which introduces branches creating alpha 1->6 linkages in starch and glycogen

Brassino steroids -one of the major group of plant growth regulators. Chemically they **tetracyclic triterpene** -polyhydroxy steroids synthesised from several sterols. Play an important role in many developmental processes.

C

C2 cycle—the photorespiratory cycle is referred to as C2 cycle as the compound 2-phosphoglycolate is a 2 carbon compound in contrast to the metabolism of 3 PGA(3-phosphoglyceric acid) which is called C 3 cycle.

CA1P 2-carboxy arabinitol-1-phosphate—an inhibitor of Rubisco formed during dark.

CABP-2'-carboxy-3-keto-D-arabinitol 1,5,-bisphosphate—an unstable six carbon intermediate in the carboxylase reaction of Rubisco.

Calmodulin (CaM)—it is a calcium binding protein which changes its conformation on binding calcium ions. When bound to calcium it can interact with many other enzymes and regulate their activities. It is one of the components of signal transduction cascade.

CAM (crassulacean acid metabolism)—the carbon dioxide fixation pathway as initially found in the members of the plant family crassulaceae. In this pathway stomata remain open during night and the initial carbondioxide fixation takes place in the night. Subsequent morning the previously fixed carbondioxide is refixed by the conventional C3 pathway by Rubisco.

cAMP phosphodiesterase—an enzyme which hydrolyses 3′,5′ cyclic AMP (adenosine monophosphate) into 5′AMP.

CAP—the structure 7- methylated Guanosine at the 5′ end of eukaryotic mRNA

Carbamylate—a reaction at the catalytic site of Rubisco between the side chain of lysine and a carbon dioxide molecule leads to the formation of carbamylate. Only then Rubisco becomes activated.

Carnitine—it is a carrier of fatty acids, helping the long chain fatty acids (14carbons and above) to be transported from cytosol to mitochondria in the form of carnitine esters. The ester is formed between fatty acid and the hydroxyl group of carnitine at the third carbon atom. Carnitine is a 7-carbon metabolite with carboxyl group at one end and positively charged nitrogen at the other end with three methyl groups. Once inside the mitochondira fatty acids once again become Co-A derivatives and metabolised by beta-oxidation. Carnitine returns to cytosol. Thus, carnitine shuttles between cytosol and mitochondria.

Cascades—the term is used to denote a sequence of steps starting from the perception of a signal till the final response is elicited in the signal transduction processes.

Caspases—they are a class of proteases which have cysteine at their active site and break thepeptide bond in a protein next to aspartate residue. An important enzyme in apoptosis.

Cataplerotic reactions—they are converse of anaplerotic reactions. These reactions drain off metabolites from a pathway.

Catecholamine—biologically active amino derivatives of catechol. Two major catecholamines are epinephrine(adrenaline) and norepinephrine(nor-adrenaline). Tyrosine is a precursor for these. Both can be classified as neurotransmitters and also as hormones.

cGMP phosphodiesterase—an enzyme which hydrolyses 3′,5′ cyclic Guanosine monophosphate into 5′ GMP.

Chalcone synthase (CHS) – first enzyme in the flavonoid pathway in plants which catalyses the reaction between three molecules of malonyl Co-A and 4-coumaroyl Co-A to give 4,2′,4′,6′ tetrahydroxy chalcone.

Cholesterol—it is a major sterol (i.e. a metabolite with a steroid ring), a 27-carbon compound. It has many biological functions and is of biomedical importance.

Cis acting factors/elements—they refer to the segments of nucleic acid which can affect only sequences lying on its own molecule either of DNA or of RNA. They interact with many regulatory molecules. The size of this cis-acting sequence elements varies from a few nucleotides to one kb.

Codon usage pattern—Many amino acids can be coded by more than one codon which is called codon degeneracy. This can vary from 2 to 6 i.e some amino acids can be coded by as many as six codons. A particular organism may have a distinct preference for the various codons to code for the same amino acid. Hence the pattern of usage of different codons may vary across the organisms.

Compartmentalization—the process of occurring of different reactions or pathways in different sub-cellular organelles.

Competing pathways—the same metabolite may be used as a precursor for more than one pathway. Such pathways are called competing pathways.

Conversion - C3 plants into C4 plants—this is one of the ambitious biotechnological goal being pursued in agriculture. The objective is to convert C3 crops into C4 crops hoping to eliminate photorespiration and thereby increase seed/grain yield.

CpG doublets-methylation—Cytosine linked to Guanine. Such pairs at specific sites undergo methylation on cytosine at the fifth position.

CpG islands-methylation—they refer to a long stretch of CpG doublets. They are found at the promoter regions of the genes. Methylation on cytosine leads to inactivation or decreased activity of the genes.

Cumulative feedback inhibition—the combined effect of inhibition of a regulatory enzyme simultaneously by several end products. Common in the biosynthetic pathways of amino acids.

Cytokinins -purine adenine derivatives. Play many roles in plant development. Exploited in tissue culture technology. Both natural and synthetic cytokinins are available.

D

Dscam(Down syndrome cell adhesion molecule) gene in Drosophila – an extreme example of alternative mRNA splicing which results in 30,186 different mRNA's

***De novo* biosynthetic pathway of purines** – entire pathway of biosynthesis of purines starting from various precursors.

Deadenylylation – removal of adenyl group from an enzyme resulting in a change in the activity (see adenylylation). Both adenylylation and deadenylylation are carried out by the same polypeptide. However, the enzyme is commonly referred to as adenylyl transferase.

Debranching enzyme–the enzyme which removes branches, that is alpha 1->6 linkages in starch and glycogen and attaches to a chain by alpha 1-> 4 linkages.

Defects in signal transduction pathways—when once signal transduction process is initiated it needs to be terminated after the required time. If the process is permanently switched on or switched off, these situations are referred to as defects in signal transduction pathways, which can lead to diseases.

DELLA domain proteins – a class of repressor proteins involved in the plant hormone gibberellin signal transduction.

Deubiquitination – removal of ubiquitin protein from the target protein.

Deuridylation – removal of uridyl group (UMP) from an enzyme which results in a change in its activity.

Dicer – an endonuclease which cleaves long double stranded RNA molecules into short RNAi (RNA interference) molecules.

Dithiol-disulfide interconversions – interconversion of 2SH (thiol group) and —S—S—(disulphide bond) in a protein or molecules like glutathione.

DNA methyl transferases – enzymes which methylate nucleotides in the DNA molecule for example cytosine.

dNDP formation – DNA requires deoxyribose sugar. The conversion of ribonucleotides to deoxyribonucleotides occurs at the level of nucleoside diphosphate. Then it is called dNDP.

Downregulated – decrease in the level of transcription of a gene.

E

EFA (essential fatty acids) - fatty acids which cannot be synthesised by mammals but had to be supplied in the diet, namely linoleic and linolenic acids. These fatty acids are further converted into more unsaturated fatty acids which can in turn be used for the synthesis of several physiologically active compounds.

EF-hand family – a structural motif present in calmodulin and also in other proteins.The EF-hand motif contains a helix-loop-helix structure looking like a thumb which is spread and forefinger of the human hand. Hence the suffix "hand".

EGF (epidermal growth factor) – it is a short polypeptide involved in cell growth, cell proliferation and differentiation.

EGF receptor – it is an enzymatic protein possessing tyrosine kinase activity (protein kinase which phosphorylates on tyrosine residues) EGF binds to this receptor to activate the enzyme.

Elongase – an enzyme in the endoplasmic reticulum which further elongates the fatty acids synthesised by the fatty acid synthase complex.

End product repression – the binding of an end product of the biosynthetic pathway to the operon inhibiting the transcription of the gene. This process regulates the enzyme amounts.

End product inhibition – the binding of an end product to an allosteric site of an enzyme decreasing the activity of the enzyme.

Endoduplication – it is a situation where in several rounds of the entire DNA of the chromosome is replicated several times as observed in the in case salivary gland chromosome of Drosophila. as a result, a giant chromosome is obtained a condition called polytene. Such chromosomes are called polytene chromosomes.

Endothelium – derived relaxing factor(EDRF) – an alternative name for nitric oxide as it is derived from endothelium.

Enhancers – cis acting sequences which helps in increasing the transcription of the gene. They can be found either upstream or downstream or even in the middle of the gene and can function in either orientation.

EPSP (5-enolpyruvylshikimate-3-phosphate)synthase – an important enzyme in the shikimate pathway leading to aromatic amino acid biosynthetic pathway. This enzyme has been exploited as target for the herbicide glyphosate (trade name "Round Up").

Erucic acid – it is 22 carbon fatty acid with double bond between 13th and 14th carbon. Present in rapeseed oil (*Brassica napus*). High levels of erucic acid are harmful to humans. However,erucic acid also has industrial applications. Accordingly, low erucic acid rapeseed (LEAR) and high erucic acid rapeseed (HEAR) verities have been developed.

Essential amino acids (EAA)–amino acids which cannot be synthesised by mammals but have to be supplied in the diet in the required amounts.

Ethylene C_2H_4 -gaseous plant hormone synthesised from methionine. Play a role in many developmental processes.

Exergonic – reactions in which energy is liberated are called exergonic reactions.

Exons– the expressed segment of nucleic acid in the matured mRNA where the genes are interrupted.

Extein–expressed part of the peptide after protein splicing.

F

Feed forward activation – the process where the substrate of an enzyme activates the enzyme to increase the rate of the reaction.

Futile cycle – in case opposing pathways that is catabolic and anabolic pathways are operating simultaneously it will result in a wastage of energy such cycles are called futile cycles. The concept can also be applied to an individual reaction where in ATP (energy) is used in one direction and formed in the reverse direction. If the reaction is operating simultaneously in both the directions, it will result in the waste of energy. Such a reaction is also called futile cycle.

G

Gibberellins – they constitute a large number of compounds produced both by plants and fungi. Chemically they are tetracyclic diterpenes. They constitute one of the major five classes of plant hormones. Play a role in many developmental processes. Exploit in tissue culture technology.

Gene amplification – increasing the number of copies of a gene.

Giant polytene chromosome (see endoduplication).

Glucagon – a short polypeptide hormone of 29 amino acids secreted by alpha cells of pancreas involved in the regulation of the glucose levels.

Gluconeogenesis – the synthesis of glucose from non-carbohydrate precursors.

Gluconeogenic amino acids – amino acids which can be catabolised and lead to the biosynthesis of glucose.

Glutamate family – glutamate, glutamine, proline and arginine.

Glutamine-oxoglutarate amidotransferase (GOGAT) – it is also called glutamate synthase and is an important enzyme in the assimilation of ammonia.

Glutathione (GSH) – it is a tripeptide—gamma glutamyl-cysteinyl-glycine— present in all organisms. Two molecules of glutathione can interact to form oxidised glutathione with a disulphide bond between the cysteine residues. Hence it's can survey redox function.

It helps in the maintaining the SH groups of the proteins in the reduced state and is also involve in the removal of toxic peroxides.

Glycogenin protein primer – it is a homodimeric protein. It has intrinsic glucocyltrasferase activity. Because of this activity it can transfer few glucose residues on to a specific tyrosine residue. The glucose units are linked by alpha 1,4 linkages. This oligomer acts as a primer for glycogen synthesis. The donor for glucose is UDP (uridine diphosphate) glucose.

Glyoxylate cycle – it is a cycle unique to plants and certain microorganisms by the help of which they can convert fatty acids/ acetate units into glucose. Animals do not posses this cycle and hence cannot convert lipid (fatty acids) into carbohydrate. This cycle is particularly important during germination of oil seeds.

Glyoxysomes – specialised microbodies in which glyoxylate cycle takes place.

Glyphosate (trade mark-"Roundup") herbicide resistant plants (see EPSP synthase) – crop varieties have been produced which are resistant to the above herbicide and they possess resistant EPSP synthase.

GOT (glutamate oxaloacetate transaminase also called aspartate amino transferase) – it is one of the important transaminases which catalyses the reversible reaction of transfer of an amino group from glutamate to oxaloacetate to form aspartate and alpha ketoglutarate. This enzyme is significant in the synthesis of above non-essential amino acids and in clinical medicine. This enzyme is also known as aspartate amino transferase.

Gout (see allopurinol) – a disease of the joint by over accumulation of uric acid and which deposited in the form of sodium urease crystals. Kidneys also affected.

G-protein coupled receptors (GPCR) – it is another name for 7TM receptors.

G-proteins – heterotrimeric proteins consisting of alpha, beta and gamma subunits playing an important role in the signal transduction process. The alpha subunit has intrinsic GTP hydrolysing activity.

GPT (glutamate pyruvate transaminase also called alanine amino transferase) – it is one of the important transaminases which catalyses the reversible reaction of transfer of an amino group from glutamate to pyruvate to form alanine and alpha ketoglutarate. This enzyme is significant in the synthesis of above non-essential amino acids and in clinical medicine. This enzyme is also known as alanine amino transferase.

Growth hormone – it is a polypeptide hormone secreted by pituitary gland which stimulates growth and development in animals and humans.

Guanylyl cyclase – the enzyme which catalyses the formation of 3′5′ cyclic GMP from GTP.

H

Histones – proteins rich in basic amino acids bound to DNA. They play an important role in regulating the chromatin structure and gene expression.

HMG CoA reductase – this is the rate limiting enzyme in cholesterol biosynthesis and is an important target for cholesterol lowering drugs.

House keeping or constitutively expressed genes – genes which are in general expressed in all cell types as they are basic to the functioning of the cells. These genes are less or non-methylated.

I

Inborn errors – genetic abnormalities manifested by birth.

Inorganic pyrophosphatase – it hydrolyses inorganic pyrophosphate into two molecules of inorganic phosphate. Inorganic pyrophosphate is produced as a product in many reactions. The action of this enzyme helps in pulling the reaction forward.

Inositol 1,4, 5-triphosphate (IP3) – it is one of the important second messengers in the signal transduction process.

Insulin – is a polypeptide consisting of two chains A (21 amino acids) and B (30 amino acids). In a larger inactive precursor form in the beta cells of pancreas. Plays an important role in carbohydrate and lipid metabolism and is of clinical significance in diabetes.

Intein – the portion of the peptide segments removed after protein splicing.

Interrupted genes – genes which have both exons and introns (see exons and introns)

Introns – the segments of RNA which are removed during splicing and do not appear in the matures mRNA.

Isocitrate lyase (ICL) – a key enzyme of glyoxylate cycle which cleaves isocitrate into glyoxylate and succinate.

Isoenzyme – multiple molecular form of a given enzyme. Isoenzymes play an important role in the regulation of many metabolic pathways.

J

Janus kinase 2(JAK 2) – one of the important protein kinases in signal transduction process. It phosphorylates tyrosine residues in protein.

Jasmonates (JA's) - Jasmonate – one of the important signalling compounds in plants. It is synthesised from alpha linolenic acid and structurally similar to mammalian prostaglandins. They play an important role in regulating many plant processes.

JAZ proteins – they are transcription repressors proteins involved in jasmonates signal transduction.

K

Ketogenic amino acids – all those amino acids are which on catabolism give rise to ketone bodies namely acetone, acetoacetate and D-beta hydroxybutyrate.

Kranz anatomy – a specific tissue organization consisting of two different cell types namely mesophyll cells and bundle sheath cells in leaves of C4 plants.

L

L8S8 – the oligomeric subunits structure of Rubisco enzyme. L8 indicates 8 large subunits and s8 indicates 8 small subunits.

Lesch-Nyhan syndrome – a genetic disorder in the male children affecting the central nervous system due to deficiency of the enzyme, hypoxanthine-guanine phosphoribosyl transferase-an enzyme in the salvage pathway of purine biosynthesis.

Lignin – most abundant aromatic polymer on earth. Provides mechanical supports to plants.

M

Malate synthase – the key enzyme of glyoxylate cycle which catalyses the reaction between glyoxylate and acetyl CoA to malate and CoA.

Malonyl- ACP (acyl carrier protein) – the three-carbon unit which repeatedly condenses with acetyl CoA during fatty acid synthesis. In this reaction one carbon of the malonyl group is lost as CO_2.

MAP (mitogen activated protein) kinases– these are a family of protein kinases phosphorylating specific serine and threonine residues in many proteins. These kinases play an important role in many physiological processes in all multicellular organisms.

Metabolons – they are multi enzyme complexes. In such complexes the product of one enzyme can easily be passed on to the next enzyme and so on.

Methylation of histones– methylation is carried out by the enzyme histone methyltransferases. Methylation regulates gene expression and chromatin structure.

Methylation of DNA – methylation is carried out by DNA methyltransferases. Methylation reduces the transcription of genes. Methylation pattern of DNA can be a characteristic feature of a cell. It is one of the epigenetic modification.

Micro RNA's (miRNA's) – short RNA molecules which regulate gene expression.

Monocistronic – an mRNA coding for only one polypeptide.

Monomeric G-proteins -these are single subunit (alpha subunit) G proteins unlike the hetrotrimeric G proteins which consists of alpha. beta and gamma subunits.These proteins also play a role in signal transduction.

mRNA editing – change of one nucleotide base into another in the matured mRNA which can result in a change in the amino acid or premature stopping of the polypeptide synthesis.

mRNA splicing– selective cutting and removal of RNA segments and joining of the remaining segments in the initially transcribed pre-mRNA to produce a mature m RNA

Multienzyme complex – a group of enzymes related to a particular pathway or a reaction remaining associated together for examples, PDH (pyruvate dehydrogenase) multi enzyme complex involved in the conversion of pyruvate to acetyl CoA and fatty acid synthase complex involved in the biosynthesis of fatty acids.

Multiple ubiquitination–addition of several ubiquitin molecules to each otheri.e in tandem on a protein molecule. First ubiquitin molecule is attached to a specific lysine residue on the target protein to be degraded. The term may also have referred to attachment of ubiquitin to several lysine residues on a protein.

N

Naturally occurring cytokinins – iPA (isopentenyl adenine), cis and trans zeatin and dihydrozeatin. Play a role in regulating many developmental prosses.

NDP (nucleoside diphosphate) kinase – this is a very important enzyme involved in the conversion of NDP's (nucleoside diphosphates) in to NTP's (nucleoside triphosphates). The enzyme does not discriminate between ribose NDP's (purines or pyrimidines) and deoxyribose NDP's (purines or pyrimidines). The donor of the phosphate groups is usually ATP though other NTP's also may be used.

N-end rule–refers to the specific amino acids at amino terminal end which determines the half-life of the protein. Proteins with different amino acids at the amino terminal end have different half-lives. For example- proteins with arginine at the amino terminal end in general have a very short half-life (approximately two minutes) while proteins with amino acids for example- methionine, glycine etc., have a greater half-life (more than 20 hours).Half-life indicates the time taken for 50% of the protein to be degraded or 50% of enzyme activity to be lost.

Nitrate reductase (NR)–the first enzyme in the assimilation of nitrate which catalysed the reduction of nitrate to nitrite, it is cytosolic enzyme.

Nitric oxide (NO) (see EDRF) – an important signalling molecule synthesise from arginine by nitric oxide synthase (NOS) it is a gaseous molecule and a free radical. Primarily involved in increasing smoot muscle contraction, vasodilation and blood flow.

Nitrite reductase (NiR) – it is the second enzyme involved in the nitrate assimilation it reduces nitrite to ammonia and localised in plastids.

NMP (nucleoside monophosphate) kinase– it is a very important enzyme involved in the conversion of NMP's (nucleoside monophosphates) either of ribose or deoxyribose into NDP's (nucleoside diphosphates). The donor of phosphate is usually ATP. NMP kinases specific for a nucleotide base exists.

NO signalling–signal transduction cascade mediated by nitric oxide. Such signalling occurs both in mammals and plants.

Non-essential amino acids (NEA) – those amino acids which can be synthesised by the mammals and need not have to be supplied in the diet.

N-terminal end signal peptide – short sequence of amino acids at amino terminal end of the polypeptide which determines the localization of the polypeptide to a specific sub cellular target sites.

Nullisomic– a genetic condition in whichboth copies of the same chromosome or missing in a diploid cell or an individual.

O

Oligomeric – the term refers to a functional protein or an enzyme with multiple sub units i.e. made up of multiple similar or different polypeptide chains.

Omega fatty acids (see essential fatty acids) –omega carbon refers to the methyl group carbon of the fatty acids and numbering of a fatty acid may be done starting from the omega carbon atom. Hence omega 3 fatty acids mean the 3rd carbon from the methyl group has a double bond. Similarly, omega 6 fatty acids mean the 6th carbon from the methyl group has a double bond.Examples for omega 3 fatty acids is alpha linolenic acid and for omega 6 fatty acids linoleic acid, gamma linolenic and arachidonic acid.

Omega oxidation – the process in which the terminal methyl group of a fatty acid is oxidized to carboxylic acid.

One carbon metabolism – reactions in which there is a transfer of one carbon units. Important metabolites involved in the metabolism are folate derivatives namely, tetrahydrofolate, 7,8, dihydrofolate and N^5-N^{10} methylene tetrahydrofolate.

Operons– in prokaryotes many mRNA's are polysistronic (i.e. specify more than one polypeptide). The unit or segment of m RNA constituting the structural genes and regulatory elements is referred to as an operon.

ORF (open reading frame) – a stretch of DNA which has the potential to be translated into a polypeptide. This stretch of DNA has a long continuous sequence of sense codons starting from the initiating codon and ending with a terminating codon.

Osmolyte – also known as compatible solutes accumulated in the cell and function in osmotic adjustment. They are usually accumulating in response to abiotic stresses like drought and salinity. There are different types of osmolytes.

Oxidative phosphorylation – the synthesis of ATP as occurs in the mitochondrial electron transport and chloroplast electron transport. The former is referred to as mitochondrial oxidative phosphorylation and the later as photosynthetic oxidative phosphorylation. This is in contrast to the synthesis of ATP (or any other energy rich compound that can be converted to ATP) in a metabolic reaction where the process is termed as substrate level phosphorylation.

Oxygenase reaction – a reaction catalysed by Rubisco where oxygen is used as a substrate. In this reaction, instead of two molecules of three phosphoglyceric acid as in carboxylase reaction of Rubisco, we get only one molecule of 3 phosphoglyceric acid and one molecule of 2 phosphoglycerates which is the two-carbon compound.

P

Passenger strand – siRNA's (small interfering RNA's) have two strands of RNA's. The two strands are separated when siRNA is unwound. The strand that is degraded is called passenger strand

and the strand that is incorporated into the RNA-induced silencing complex (RISC) is called guide strand.

Pasteur effect – inhibition of fermentation pathways of glycolysis in the presents of oxygen.

Pathogenesis related proteins (PR) – these are specific plant defence proteins induced in response to microbial pathogens.

PEST sequences – the letters represent the single digit code for the amino acids proline, glutamic acid, serine and threonine respectively. Proteins having segments which are rich in these amino acids are more susceptible to ubiquitin mediated protein degradation.

PH domain (Pleckstrin homology) – it is a domain of about 120 amino acids situated at the amino terminal end of the enzyme phospholipase C which plays an important role in the signal transduction of phosphoinositide cascade. This domain was first found in platelet and leukocyte C kinase substrate protein.

Phenolics – a major category of secondary plant metabolites. The basic unit of phenolic compound is an aromatic ring with one or more hydroxyl groups attached to it.

Phenylalanine ammonia lyase – it is first enzyme involved in the biosynthesis of numerous phenolic compounds. It catalysed the removal of the amino group from phenylalanine to form tras-cinnamic acid which further undergoes metabolism.

Phosphoenolpyruvate (PEP) carboxylase – it is the initial CO_2 fixation enzyme in C4 plants. In this process PEP is converted to oxaloacetate.

Phosphoenolpyruvate carboxykinase – this an important enzyme in the reversal of glycolysis i.e. gluconeogenesis. It decarboxylases oxaloacetate to give PEP thus bypassing the irreversible reaction of PEP to pyruvate. Oxaloacetate can be formed from pyruvate by pyruvate carboxylase reaction.

Phosphoinositide cascade – signal transduction process involved mediated by phosfatydyl inositol bisphosphate (PIP2)

Phosphoprotein phosphatase – these enzymes remove phosphate groups from proteins. The phosphate groups were added to the proteins by the enzymes protein kinases.

Phosphorolytic cleavage – it denotes degradation of starch and glycogen using inorganic phosphate with the formation glucose 1 phosphate.

Photoperiodism – the differential response of the plants to changes in the duration of day and night.

Photorespiration – the process where in plants consume oxygen in the presence of light and produce carbon dioxide. The stating metabolite for this process is 2 phosphoglycerates formed in the oxygenase reaction catalysed by Rubisco. The entire pathway takes place in different subcellular organelles in the leaves. It is considered apparently as a wasteful process consuming energy.

Phytochelatins – they are metal binding short polypeptides of a different sizes and of the general composition (gamma Glutamic acid- Cysteine)$_n$Glycine (n varies from 2 to 11) they are synthesised to response to high concentration of certain metals.

Phytochrome – it is a photoreceptor plant pigment. Chemically it is the biliprotein consisting of a chromophore and an apoprotein. It exists in two interconvertible forms P_R(red light approximately 660nm, absorbing form) and P_{FR}(far-red light approximately 730nm, absorbing form). This photoreceptor is involved in regulating many developmental processes.

Ploidy levels – Ploidy is the number of sets of chromosomes in acell. Gametes are haploid (only one set) and somatic cells are diploid (two sets). In plants we may have tetraploid and hexaploidy etc.,

Poly A tail at the 3 prime end – eukaryotic mRNA's have a stretch of adenine residues (AMP) at the 3 prime end of the mRNA. The stretch can be as long as 200 adenine residues. This addition is catalysed by poly A polymerase. (Histone mRNA's lack poly A tails)

Polyamines – small metabolites having two or more amino groups. Most common poly amines are putrescine, spermine and spermidine, cadaverine and thiospermine. They are present both in plants and animals and regulate many physiological processes.Being positively charged they bind to DNA.

Polycistronic mRNA – characteristic features of prokaryotes where in the m RNA specifies more than one poly peptide.

Polyubiquitin(see multiubiquitin).

PPi (inorganic pyrophosphate) dependent PFK – unlike other PFK's (phosphofructokinases) namely PFK1 and PFK2 which use ATP as the donor of phosphate group this enzyme uses inorganic pyrophosphate as the phosphate donor to form fructose 1,6 bisphosphates.

Primary messenger – the initial signal or the ligand which imitate the signal transduction cascade is called as primary messenger.

Pro-systemin – systemin is a short polypeptide of 18 amino acids formed as a result of proteolytic cleavage of its larger precursor pro systemin in response to wounding. It activates the jasmonic acid biosynthetic pathway. It is a defence response against insect attack.

Proteasome – large multi protein complexes involved in protein degradation. Ubiquitin mediated protein degradation is accomplished by proteasome.

Protein kinases – they are a family of enzymes phosphorylating various proteins usually at serine, threonine and tyrosine residues. This is an important covalent modification involved in the regulation of the activity of many enzymes.

Protein phosphatases – they are a family of enzymes which remove phosphate groups from proteins which have been added by protein kinases. This process is called dephosphorylation and is involved in the regulation of many enzyme activities. Phosphorylation – dephosphorisation is a reversible covalent modification on proteins.

Protein sorting or protein trafficking or protein targeting – the process by which protein synthesised by cytosol are directed to different intracellular sites.

Protein splicing – the cutting and removal of unwanted peptide segments and joining of the remaining required segments to form a functional polypeptide.

Protein turnover – the process of constant synthesis and degradation of proteins.

Proteome – the sum total of proteins in a cell or in an individual organism.

PUFA (poly unsaturated fatty acids) – fatty acids with multiple double bonds (see omega fatty acids)

Pyrophosphatases – these are enzymes which hydrolyse inorganic pyrophosphate liberating two molecules of inorganic phosphates. The enzyme by catalysing this reaction pulls the reaction forward.

Pyruvate family of amino acids – they constituteisoleucine, leucine and valine and alanine.

Pyruvate inorganic phosphate dikinase – also known as PPDK (pyruvate-orthophosphate dikinase) it is a key enzyme in C4 plants which catalyses the conversion of pyruvate into PEP in mesophyll cells.

R

R2C2 – it is an inactive tetrameric structure of protein kinase A. when cAMP binds to R2C2 the catalytic chains C2 are liberated.

Receptor tyrosine kinases – membrane protein receptors which posses' intrinsic tyrosine kinase activity which phosphorylates proteins on a tyrosine residue.

Reciprocal regulation (see futile cycles) – opposing pathways are not simultaneously active/operative.

Reducing power – organic compounds which can donate hydrogen atoms. Examples, $NADH+H^{+,}$ $NADPH+H^{+,}$ $FADH_2$ etc.,

Reductive pentose phosphate pathway – refers to Calvin cycle in contrast to oxidative pentose phosphate pathway (also called HMP pathway or Shunt) which is involved in oxidation of glucose 6 phosphate. Many reactions are common in both the pathways. However, the functions of the pathways are different.

Ribozymes – these are RNA molecules possessing catalytic activity.

Ribulose-1,5, -bisphosphate (RuBP) – the acceptors of carbon di oxide in the first reaction of Calvin cycle catalysed by Rubisco.

RISC (RNA induced silencing complex) – it is ribonucleoprotein complex. The complex has a nuclease and siRNA. The complex cleaves m RNA's which are complementary to guide strand (see passenger strand).

RNA interference (RNAi) – this is the process in which large precursor double stranded RNA molecules are cleaved to short

RNA molecules which are complementary to various mRNA's. As a result of the complementarity either mRNA translation is inhibited or mRNA is degraded.

RNA silencing – the process by which non-coding RNA's inhibit the gene transcription.

RNA splicing – the process by which from the precursor m RNA introns are removed and exons are joined to give a matured mRNA.

Rubisco – ribulose-1,5, -bisphosphate carboxylase /oxygenase the first enzyme in the Calvin cycle responsible for fixing carbon dioxide. A bifunctional enzyme which can also use oxygen as a substrate leading to photorespiration. It is the mostabundant enzyme/protein on earth.

Rubisco activase – an enzyme involved in the removal of the inhibitor CA1P from the Rubisco and activates it.

S

SA (Salicylic acid) – chemically it is 2-hydroxy benzenecarboxylic acid involved primarily in the induction of flowering, thermogenesis, disease resistance - localized and systemicacquired resistance.

SAR(systemic acquired resistance) – the phenomena where the resistance developed by plants against pathogens is not confined only to the part of infection but also transmitted to sites distant to the site of infection.

S-adenosyl methionine(SAM) – also called active methionine and is a donor of methyl groups in several important reactions.

Salvage pathway of purines – the synthesis of purine nucleotides from pre-formed purine bases in contrast to de novo pathway of purines where the purines are synthesized in an elaborate set of reactions starting from various precursors.

Second messengers – regulatory molecule synthesized in the cell in response to the binding of a signal/ligand which is called as the primary messenger.

Serine family Amino acids- they constitute serine, glycine and cysteine.

Serpins – serine protease inhibitors

S-glutathionylation – it is one of the important post translation modification where in glutathione is added to cysteine residue in proteins. Plays a role many important physiological processes.

SH – it refers to Src (sarcoma) homology. This is a peptide domain which binds to phosphotyrosine reside in a protein.

SH3 domains – these are peptide domains which bind to proline rich segments of polypeptides.

Signal peptidase – an enzyme which cleaves the signal sequence in a protein which is responsible for protein targeting.

Signal sequence – a stretch of amino acids present either at the amino terminal end or carboxyl terminal end or within the protein which is responsible for protein targeting.

Sildenafil citrate – it is a drug whose trade name is Viagra. This drug stimulates penile erections by enhancing signalling through nitric oxide pathway.

Silencers – short segments of DNA that can inhibit the expression of the genes.

S-nitrosylation – an important reversible post translation modification of cysteine residues of proteins by nitric oxide. This modification plays an important role in many physiological process. Many transcription factors undergoes this modifications and regulates the transcription of genes.

Sodium glutamate – a flavour enhancer in Chinese foods like noodles.Its taste is referred to as Umami.

Somatotropin (see growth factor)

Starch – reserve carbohydrate in various plant organs.

START domains – these are polypeptide domains which are present in ABA receptors START is an acronym for “star related lipid transfer”. These domains are usually present at the C – terminal end and is involved in mediating the hormone responses.

STAT 5 – STAT is an acronym for signal transducers and activators of transcription. They possess SH2 domains STAT 5 as dimer binds to specific sites on DNA and regulates gene expression. It is an important component of signal transduction pathway.

Statins – they are a family of a drug's use to lowering the cholesterol level. They inhibit the key regulatory enzyme HMG-CoA reductase in the cholesterol biosynthesis.

Stringent response – when bacteria are starved for amino acid it leads to reduction of r RNA synthesis and thereby ribosomes and also tRNA synthesis. This process is mediated by the accumulation of guanosine tetra (ppGpp) and penta phosphates (pppGpp).

SUMO (small ubiquitin related modifier) – the addition of SUMO is one of the important covalent modification of histones on a specific lysine residue. The process is called SUMOylating and is involved in the regulation of chromatin structure and gene expression.

synthetic cytokinins - kinetin (6-furfurylamino purine), 6-benzyladenine. Exploited in tissue culture technology.

T

Transaminases (see **GOT** and **GPT**).

Triose phosphate translocator (TPT) – This is an integral membrane protein found in the inner membrane of chloroplasts. It exports triose phosphate (Dihydroxyacetone phosphate) in exchange for inorganic phosphate. This protein is responsible for exporting all the carbohydrate produced in photosynthesis by plants. Hence genetic engineering for overexpression for this protein is attempted.

U

Ubiquitin – it is a short protein of 76 amino acids and is highly conserved among all categories of organisms. It is covalently attached to side chain of lysine and mediates protein **degradation** (see multiubiqutination).

UDPG (uridine di phosphate glucose) – it is sugar nucleotide which donates glucose in glycogen biosynthesis. Rarely in starch biosynthesis.

Upregulated – increase in the transcription of genes.

Urea cycle – it is a pathway of nitrogen/ammonia excretion in ureotelic organisms. Majority of the terrestrial organisms including humans are ureotelic.

Ureides (allantoin and allantoic acid) – these are degradative products of purines. They constitute an important nitrogen transport mechanism in nodulated legumes.

Uridylation – the process of addition of UMP to specific tyrosine on an enzyme/protein catalysed by the enzyme uridylyl transferase. This is one of the reversible post translational covalent modification.

V

Vasopressin – it is a short peptide hormone of nine amino acids synthesised by posterior pituitary.

Z

Zymogen activation – many enzymes/proteins are initially synthesized in a longer inactive form. These precursors either have the suffix "gen" or prefix "pro". These precursors are then proteolytically processed to give the active form of the enzyme/protein.

2-phosphoglycolate – first metabolite in the photorespiratory cycle formed as one of the products of oxygenase reaction of the enzyme Rubisco.

3PGA — 3-phosphoglyceric acid, a product of the first reaction in the Calvin cycle catalysed by the enzyme Rubisco.

7TM receptors — a major class of membrane protein receptors involved in signal transduction. TM indicating the transmembrane nature and 7 indicating that the protein coils 7 times in the membrane.

PART - I

BASIC CONCEPTS/TYPES OF METABOLIC REGULATION INVOLVED IN THE REGULATION OF ENZYME ACTIVITY/AMOUNTS

1

AN OUTLINE OF VARIOUS TYPES OF METABOLIC REGULATION

Changes in substrate/metabolites concentration

Students do study in a basic enzymology course about the increase in rate of reaction with increase in substrate concentration, hyperbolic curve i.e. Michaelis Menten (MM) kinetics, the concept of K_m and V_{max}, and different types of enzyme inhibition. However *in vivo* substrate concentrations are not usually so high but found in low concentrations to exhibit MM kinetics. Hence changes in substrate concentration is not the main determining factor for the regulation of enzymatic activity of enzymes exhibiting hyperbolic curves. However, for allosteric enzymes -which show sigmoidal substrate saturation kinetics, variation in substrate concentration can play a significant role. The activity of the allosteric enzymes is also significantly affected by changes in metabolites, other than substrates -also called effectors, modulators, ligands, regulators- which can either activate or inhibit the enzymes. The binding of these metabolites to allosteric enzymes is non-covalent and reversible. There are as many binding sites as there are activators and inhibitors. The final activity of these allosteric enzymes can be a result of a ratio of the concentration of their activators and inhibitors. Many rate determining enzymes are allosteric enzymes. These enzymes are usually present at the first step in the metabolic pathway or at the first step of the branched metabolic pathway. These reactions are called rate limiting reactions/steps in the metabolic pathway and the concerned enzymes as pacemaker enzymes. This type of inhibition is also called Feed Back Inhibition (FBI) Feedback regulation or end product inhibition or end product repression & Feedback repression (FBR)- This mode of regulation is most common among all organisms. In this case as mentioned above the end product of the main pathway inhibits the enzyme of the first reaction in the pathway or the end product inhibits the enzyme of the first reaction in the branched pathway. The end product can allosterically inhibit the activity of the enzyme and/or repress the synthesis of the enzyme acting on genetic level. The same enzyme can be inhibited by more than one end product. Sometime this inhibition and repression is selective for a specific isoenzyme. Sometimes the substrate can activate the enzyme acting on it, a phenomenon called feed forward activation. The term feedback regulation is often used in a broader sense. For example, when the intake of cholesterol in diet increases it results in a decrease of cholesterol biosynthesis. However, the dietary cholesterol has no effect on the rate limiting enzyme hydroxymethyl glutaryl CoA reductase. But when cholesterol is being endogenously synthesised the above rate limiting enzyme is inhibited by cholesterol. In general, the terms FBI and feedback regulation are used interchangeably. Inhibition of enzyme activity can also occur with non-allosteric enzymes.

Changes in pH

This aspect too students study in the basic enzymology course and become familiar with various types of curves (indicating how activity of different enzymes are effected) that can be obtained with alterations in pH. However, the enigma is optimum pH of an enzyme need not be the same as that of intracellular environment and thousands of enzymes are functioning at the same intracellular pH! However, any change in the intracellular pH can affect the activity of the enzymes, e.g. regulation of the activity of Calvin cycle enzymes in the chloroplast by an increase in pH towards alkaline condition.

(A) Adenylate energy charge- It is defined as "the mole fraction of ATP plus half the mole fraction of ADP". So it depends on the relative amounts of ATP, ADP and AMP.

$$\text{Adenylate energy Charge} = \frac{[\text{ATP}] + 1/2\,[\text{ADP}]}{[\text{ATP}] + [\text{ADP}] + [\text{AMP}]}$$

If all the adenylates are present as ATP, then the adenylate energy charge is equal to 1 (one). If all adenylates are present as AMP, then the adenylate energy charge is equal to 0 (zero). In normal cells the energy charge usually varies from 0.70 to 0.95 (these values may be quoted slightly differently in different text books. What is given is the entire range found). Usually when the energy charge is high, exceeds 0.80 anabolism takes priority. When the energy charge is low say below 7.0 catabolism takes priority. *E.coli* cells die if the energy charge is less than 0.50.

(B) Phosphorylation Potential- Another index of energy status often discussed is phosphorylation potential which is defined as the ratio of concentration of ATP to concentration of ADP plus Pi (inorganic phosphate).

Phosphorylation Potential = [ATP]/[ADP] + [Pi]

Though casually we talk about only anabolic pathways and catabolic pathways, it is more right to talk about another category- amphibolic pathways. A pathway, whose intermediates can be used to synthesise some other metabolites is termed as an amphibolic pathway. For example Glycolysis and TCA (Tricarboxylic acid cycle) strictly speaking should not be referred to as a catabolic pathway but as amphibolic pathway as some of its intermediates can be used for the synthesis of other compounds. In fact, precursors for the synthesis of the large number of secondary compounds are supplied from primary metabolic pathways.

Reversible covalent modifications of enzymes/non-enzymatic proteins

(A) Some examples

Enzymes at certain specific amino acid residues can undergo many reversible covalent modifications i.e. covalent attachment of some functional group. As a result of the attachment, some enzymes becomes active and some enzymes become inactive. In other words, for some enzymes to be reactivated the covalent group attached has to be removed. Some important reversible covalent modifications are – phosphorylation -dephosphorylation; adenylation-deadenylation; uridylation-deuridylation; addition and removal of ADP ribosylation. An important example of non-enzymatic proteins reversibly getting covalently modified is that of nuclear proteins (histones). Histones can be reversibly methylated, phosporylated, acetylated, ubiquitinylated, sumoylated and ADP-ribosylated. These reversible covalent modifications are known to regulate gene expression and hence regulation of enzyme amounts rather the regulation of the activity of the existing enzymes as mentioned above. Among the post translational modifications, phosphorylation seems to be the most prevalent in eukaryotes. Nearly one-third of proteins may be regulated by reversible phosphorylation. Further it should be noted that a protein can add either only one or many phosphate groups at different positions in the polypeptide. Serine, threonine and tyrosine are the most common amino acids in a protein which get phosphorylated. Histidine and aspartate also may be involved. Nearly one-third of all eukaryotic proteins are supposed to be phosphorylated. Nitric oxide (NO)-dependent S-nitrosylation of cysteine residue in proteins is a key post-translational mechanism controlling the functions of transcription factors and several other proteins thereby regulating several pathways. NO is a signaling molecule both in plants and mammals (See signal transduction).

The functional groups, phosphate and adenyl, uridyl, methyl, acetyl, and ADP ribosylation arise from ATP, UTP, SAM (S-adenosyl methionine), acetyl CoA and NAD^+ respectively. Ubiquitin is a polypeptide.

(B) Interconversion of dithiol (2SH) and disulfide (-S-S-)

The amino acid cysteine has a sulfahydryl (thiol) group (-SH) in its side chain. This can interact with another such group in a protein molecule and get converted into disulfide and vice versa. Some enzymes are active in thiol form and some enzymes in disulfide form. The regulation of Calvin cycle enzymes mentioned above due to changes in pH is followed by dithiol-disulfide interconversions in different enzymes mediated by another protein called Thioredoxin. Further there are also enzymes, protein disulfide isomerases and other enzymes which can carry out the above interconversions affecting the activity of enzymes and even metabolites like Glutathione.

Zymogen activation

A large number of both enzymatic and non-enzymatic proteins are first synthesised as larger inactive polypeptide precursors and are subsequently cleaved in a specific manner by specific proteases to liberate the active form of the enzyme/non-enzymatic protein. They are called as Proenzymes or zymogens. The process is also referred to as zymogen activation. Some common examples of zymogen activation of enzymes are—pepsinogen to pepsin; trypsinogen to trypsin; chymotrypsinogen to chymotrypsin and that of polypeptide hormone is proinsulin to insulin (Details are discussed in the text). In addition to the above mentioned some important examples are formation of neuropeptides, formation of active endocrine hormones, formation of blood clotting factors (e.g. plasminogen to plasmin; prothrombin to thrombin; fibrinogen to fibrin) and other important enzymes like procollagenase, procaspases, procarboxy peptidases. Collagenase and caspases are known to play an important role in differentiation and programmed cell death/apoptosis respectively.

Students should keep in mind that zymogen activation is an irreversible process/modification unlike "A" and "B". However, the proteases formed above—their activity can be inhibited by specific inhibitors which are also small proteins e.g. trypsin inhibitors. In general, they are called as protease inhibitors. Protease inhibitors which specifically inhibit proteases which have serine at their active site are called serpins e.g. antitrypsin and antithrombin.

However, students should keep in mind that:

(a) Cumulatively different proteins undergo a very large number of post-translational modifications. Over 500 modifications have been found (both enzymatic and non-enzymatic proteins combined).

(b) The same protein can undergo more than one post-translational modification.

(c) Not only proteins even bases/nucleotides in nucleic acids undergo many covalent modifications which play an important role in maintaining their structure and function and gene expression. All these can ultimately affect the total enzyme amounts in the cell.

Oligomerisation status

As enzymes can have multiple subunits and exists as dimers, tetramers, octamers etc, their activity may depend on the oligomerisation status i.e. less or inactive under a specific oligomeric status and more active when existing in a different oligomeric status i.e. less active or inactive under a specific oligomeric status and more active or fully active when existing in a different oligomeric status.

Regulation by compartmentalisation

Metabolic pathways may be compartmentalised for independent regulation. For example, Beta oxidation of fatty acids can take place in mitochondria (in animal systems) and fatty acid synthesis takes place in cytosol. In a cell there can be many compartments *viz.* mitochondria, cytosol, chloroplast, peroxisomes, glyoxysomes, vacuoles, lysosomes, nucleus, endoplasmic reticulum, Golgi apparatus. The entire cell is enclosed by a cell/plasma membrane and in some organisms cell membrane is further covered by a cell wall. Each sub-cellular structure can in turn be further compartmentalised. For example, mitochondria has an outer membrane, outer matrix, inner membrane and inner matrix. The transport of both macromolecules and metabolites between the organelles need to be regulated.

Regulation by isoenzymes

Isoenzymes or isozymes are multiple molecular forms of the same enzyme. They catalyze the same reaction, have the same E.C. number but differ in their kinetic and regulatory properties, amino acid sequence, isoelectric point, electrophoretic mobility and immunological properties. Many enzymes occur as isoenzymes and are involved in the regulation of different metabolic pathways both in prokaryotes and eukaryotes. Different isoenzymes may be present during differentiation, growth and development and in different organelles, tissues, and under different environmental conditions.

Concept of futile/substrate cycles and Reciprocal regulation

Suppose an ATP utilising reaction and an ATP producing reactions are operating simultaneously, it only results in the net consumption of energy i.e. only a waste of energy. If it happens it is referred to as a futile cycle or substrate cycle. Hence both reactions should not operate simultaneously. When ATP utilising reaction is active, ATP forming reaction should be inactive and vice versa. Accordingly, the enzymes of the respective reaction must be regulated. At pathway level, anabolic pathways and catabolic pathways should not operate simultaneously. If glycolysis—glucose breakdown- and gluconeogenesis—synthesis of glucose take place simultaneously, it results in a futile cycle. Hence when glycolysis is active gluconeogenesis should be inactive and vice versa. Similarly, with fatty acid oxidation and fatty acid synthesis. Accordingly, the rate limiting enzymes are regulated. Such regulation is called reciprocal regulation of metabolic pathways.

Metabolism is interconnected

Metabolic pathways are interconnected. That is, if there is a disturbance in one metabolic pathway other pathways connected to it also will be disturbed. Alternatively, if a biosynthetic pathway is interconnected all other pathways concerned with it will also have to be operative simultaneously. For example, Purine and Pyrimidine biosynthesis requires several precursors, for example, amino acids glutamine, aspartate and glycine; sugars ribose and deoxy ribose and energy in the form of ATP. Therefore, to provide the above precursors amino acid metabolism and carbohydrate metabolism have to be simultaneously be active. Then only purine and pyrimidine nucleotide biosynthesis can take place. Similarly, for the secondary metabolism to be active i.e. synthesis of various secondary metabolites, primary metabolic pathways also have to be active as they have to provide precursors, energy and reducing power for the biosynthesis of secondary metabolites.

Regulation by cell signaling

Also called signal transduction is the process by which signals from external environment are transmitted into the cell's interior.

This process is involved both in regulation of enzyme activity and also enzyme/protein amounts. It is better studied in animal/human systems than in plants.

Regulation at the levels of chromatin organisation, transcription, post-transcription, translation, post-translation and by protein targeting and protein degradation can regulate the enzyme amounts (Details discussed in text).

Both endogenous and exogenous ligands can act at genetic level and regulate both enzyme/protein amounts and activity. This phenomenon is called signal transduction (Discussed as operating in both mammals and plants).

Metabolic control analysis (MCA)

A mathematical study of the flux of substrates and products through the enzymes of a biochemical pathway is defined as Metabolic Control analysis. By an understanding of the mathematical treatment of the metabolic pathways we can understand how to manipulate the flux and thereby the concentration of metabolites. By applying MCA, it is possible to identify the steps that should be modified to achieve a successful alteration of flux or metabolite concentration in pathways of biotechnological relevance.

Part - II

Regulation of Pathways in Carbohydrate metabolism

2

Glycolysis

General Introduction

It is quite common to start with glycolysis while discussing carbohydrate metabolism and is discussed starting with Glucose as the beginning of the glycolytic pathway. It is one of the most fundamental pathway with ubiquitous distribution. Depending on the situation and system, Glucose may be available as free glucose or it can arise from the breakdown of starch, glycogen, sucrose, maltose and all other sugars which contain glucose. Sugars other than Glucose also can enter Glycolytic pathway. Starch and glycogen can give rise to both free glucose and also to Glucose -6-Phosphate on their breakdown. Glucose can also be synthesised from other precursors like lactate, glycerol or from selective group of amino acids called gluconeogenic amino acids. Plants and some microorganisms can also synthesize glucose from fatty acids because of the presence of another cycle called Glyoxylate cycle. All photosynthetic organisms can synthesise glucose from photosynthetic/Calvin cycle.

Glucose arising from any of the above ways can enter glycolysis and get converted into pyruvate through a series of ten reactions. Beyond pyruvate metabolism is decided by the availability or non-availability of oxygen. Till pyruvate formation oxygen availability is not an influencing factor.

The overall reaction for glycolysis is:

$$\text{Glucose} + 2\,NAD^+ + 2\,ADP + Pi \rightarrow 2NADH + 2\text{ pyruvate} + 2\,H_2O$$

When oxygen is deficient i.e. under anaerobic conditions pyruvate may be converted to lactate or in some microorganisms and even in plants pyruvate may be converted finally to ethanol via acetaldehyde intermediate. Plants also may convert pyruvate to lactate. Lactate formed in muscle is exported to liver, converted to glucose and supplied back to the muscle.

When oxygen is sufficiently available i.e. under aerobic conditions pyruvate enters into mitochondria and metabolised by TCA (Tri carboxylic acid) cycle also called Citric acid or Krebs cycle. Inhibition of anaerobic degradation of Glucose by providing oxygen is referred to as Pasteur effect. In some special cases, even in the presence of oxygen pyruvate may be converted to lactate i.e. Pasteur effect is lost. For example, in some tumor cells and also in normal retinal cells. They produce a different isoenzyme of Hexokinase which is insensitive to feedback inhibition by Glucose-6-Phosphate.

Regulation

(a) General Details

Regulation of glycolysis is important as it produces energy and various end products of biotechnological importance in different organisms and the intermediates are diverted for the synthesis of several other useful compounds which in turn may be used for the synthesis of several other molecules.

(b) Regulation - General details

Usually three rate limiting steps where glycolysis is regulated are discussed.

(a) Hexokinase (HK),

(b) Phosphofructokinase -1 (PFK-1),

(c) Pyruvate kinase (PK). However there is also a fourth step which can be included;

(d) Phospho glyceromutase. The first three reactions are irreversible and exergonic i.e. release considerable energy. Of all the three PFK-1 is the major controlling point. Under aerobic conditions pyruvate is transported into mitochondria to enter TCA cycle. However, under anaerobic conditions, pyruvate is reduced to lactate by lactate dehydrogenase (LDH) or is converted to ethanol by the successive reactions of Pyruvate decarboxylase and Alcohol dehydrogenase.

We use the notation PFK-1 because, PFK-1 catalyses the reaction:

Fructose-6-phosphate + ATP → Fructose 1,6, Bisphosphate + ADP

There is another phosphofructokinase, PFK-2 which catalyzes the reaction-

Fructose-phosphate + ATP → Fructose 2, 6, Bisphosphate + ADP

Similarly, we have Fructose Bisphosphatase 1 and 2, given the notation, FBPase-1 and FBPase-2. They catalyze the hydrolysis of Fructose 1,6, Bisphosphate and Fructose 2, 6, Bisphosphate respectively liberating Fructose-phosphate and inorganic phosphate. Both the kinase and phosphatase activities are found/reside on the same polypeptide i.e. it is a bifunctional polypeptide. In addition to PFK-1 and 2 there also exists in some organisms e.g. plants and protists -PPi (inorganic pyrophosphate) dependent PFK. This enzyme uses PPi as the donor of phosphate groups in contrast to ATP by PFK-1. This enzyme also can be rate limiting.

Though it is common to refer glycolysis as a catabolism of glucose or a catabolic pathway giving energy, it is also right to consider it as an amphibolic pathway as many intermediates of glycolysis are used in the synthesis of other metabolites. Few important examples are:

G6P, F6P and Glyceraldehyde 3-Phosphate (GA3P) can enter oxidative pentose phosphate (OPP) pathway.The latter two metabolites can also enter glycolysis from OPP pathway; Dihydroxyacetone Phosphate (DHAP) can be diverted to lipid synthesis and become a component of lipids; PEP is a precursor for aromatic amino acid biosynthesis which are in turn used for the synthesis of numerous phenolics particularly lignin in plants; 3PGA is a precursor of serine amino acid which in turn can be used for the synthesis of other amino acids; Pyruvate by transamination reaction can form alanine. Pyruvate is also a precursor for lactate and ethanol and many other important biotechnological products. Hence glycolysis is interconnected to several other pathways, making its regulation very important. (see discussion under TCA cycle)

(c) Regulatory details of different enzymes

(a) HK- HK in humans can exist as four isoenzymes. HK I to HK IV. HK IV is also called glucokinase which is not inhibited by Glucose-6-phosphate. While glucokinase phosphorylates glucose, glucose-6-phosphatase hydrolyses Glucose-6-phosphate liberating glucose and inorganic phosphate. These two enzymes are transcriptionally activated by glucose and the hormone glucagon resulting in higher production of glucose. Otherwise HK in general is inhibited by Glucose-6-phosphate and activated by inorganic phosphate. As PFK-1 is the major control point, when PFK-1 is inhibited it leads to the accumulation of Glucose-6-phosphate which in turn inhibits HK. HK phosphorylates not only glucose but also other hexoses like fructose etc. with different affinities and they also can enter glycolysis. Instead of free glucose when glycogen acts as a source, then we get directly Glucose -1-Phospshate which is converted to Glucose -6-Phosphate. Hence HK is dispensable and its regulation is not involved. PFK-1 only remains the major controlling point. Starch also is degraded by phosphorolytic pathway.

(b) PFK—1

(i) Allosteric Regulation and Regulation by Energy Status

PFK-1 is allosterically activated by F2, 6,BP, AMP, ADP (high levels of AMP and ADP indicates low energy status and thus a need exists for the production of more energy via glycolysis). Inorganic phosphate, ammonium ions and F-6-P also help in activating the enzyme. Activation by the substrate F-6-P is termed as feed forward activation.

PFK-1 is allosterically inhibited by protons. This inhibition prevents excess accumulation of lactate under anaerobic conditions. PFK-1 is allosterically inhibited by ATP and citrate. High ATP levels indicate high levels of energy availability and hence no need for more energy production. Similarly, high levels of citrate indicate more flow of pyruvate from Glycolysis into TCA cycle. Hence these two situations inhibit glycolysis by inhibiting PFK-1. (Energy is produced in the form of ATP and GTP. While both can be formed during substrate level phosphorylation i.e. during a metabolic reaction, ATP is formed only in mitochondria as a result of oxidative phosphorylation. In case of photosynthetic organisms ATP is also produced during Photosynthetic oxidative phosphorylation in chloroplasts/membranes. Other NTP's like TTP, UTP are formed during metabolic pathways and are considered equivalent to ATP as they can be converted to ATP by NDP kinase enzyme (Nucleoside diphosphate kinase) which catalyzes the reaction, TTP/UTP + ADP $\rightarrow$ ATP +TDP/UDP. Other NTP's also can participate in this reaction. Plants may have a different system of regulation of PFK-1. It is activated by inorganic phosphate (Pi) and strongly inhibited by Phosphoenol puruvate (PEP). Other metabolites in the glycolytic pathway also inhibit PFK-1 to varying extents.

F2,6, BP levels are regulated by the relative activities of PFK-2 and F2,6, BPase.

(ii) Covalent/Hormonal Regulation of F2, 6, BP levels

PFK-2 is stimulated by being phosphorylated mediated by cAMP and the hormone epinephrine in case of muscle resulting in stimulation of glycolysis. However, in case of liver PFK-2 is inhibited by glucagon. Glucagon induces phosphorylation of both FBPase-2 and PFK-2 mediated by cAMP. As a result, PFK-2 becomes inactive and FBPase-2 becomes active. Therefore, glucagon decreases F2,6, BP levels and inhibit glycolysis. How epinephrine, glucagon and other hormones act is discussed under signal transduction. In some animals yet another mode of regulation may be present. The pentose phosphate -Xylulose-5-phosphate activates- a phosphoprotein phosphatase 2A-which dephosphorylates the bifunctional PFK-2/FBPase-2 enzyme parallelly. While dephosphorylation activates PFK-2, dephosphorylation inhibits FBPase-2. Therefore, it results in the increase in the concentration of F2,6, BP and stimulation of glycolysis and inhibition of gluconeogenesis. Simultaneous operation of glycolysis and gluconeogenesis results in a futile cycle.

(c) Pyruvate kinase (PK)

Pyruvate kinase exhibits (i) allosteric regulation, (ii) regulation by covalent modification by phosphorylation mediated by hormones and (iii) by isoenzymic control in case of animal and human systems.

(i) PK is allosterically activated by F1,6, BP, ADP, AMP and PEP. It is allosterically inhibited by fatty acids, acetyl CoA, ATP and alanine.

(ii) PK is also inhibited by modification by phosphorylation mediated by cAMP. Here the hormone involved is vasopressin which increases calcium concentration and subsequently leads to phosphorylation of PK and its inhibition. Glucagon also can cause phosphorylation and inhibit PK. Pyruvate kinase inactivation by phosphorylation induced by hormones glucagon and vasopressin inhibits glycolysis and stimulates gluconeogenesis.

(iii) PK has two different isoenzymes. L-type in liver and M-type in muscle and brain. The activity of L-form is reduced by phosphorylation. Up to three isoenzymes can be found in higher animals. In plants pyruvate kinase is inhibited by citrate.

(d) Phosphoglyceromutase and the role of 2,3, BPG

The enzyme catalyzes the reversible reaction: 3-phosphoglycerate $\rightarrow$ 2- phosphoglycerate

Minute amounts of 2,3-Bisphosphoglycerate (2,3-BPG) are required for the enzyme to be in active state. If this metabolite is not bound, then the enzyme is not active. 2, 3-BPG is formed from 1,3-BPG by the action of the enzyme Bisphophoglycerate mutase.

(ii) Further 2,3-BPG plays a role in oxygen delivery. Adaptation at high altitudes involves increase in the amount of hemoglobin per erythrocyte and number of erythrocytes. 2,3-BPG is synthesized within the erythrocyte. High concentration of 2,3-BPG decreases oxygen binding affinity to hemoglobin and thus delivers more oxygen to tissues.

(e) LDH-isoenzymic control

This enzyme under anaerobic conditions reduces pyruvate to lactate as follows:

Pyruvate + NADH+H^+ $\leftrightarrow$ Lactate + NAD^+. This reaction helps to regenerate NAD^+ for the glycolysis to continue.

LDH is a tetramer (4 sub-units) consisting of two different types of sub-units designated "M" for Muscle, also designated as "A" and "H" for Heart, also designated as "B". In all there exists five isoenzymes or five different tetramers *viz.* H4, H3M, H2M2, HM3 and M4. They can also be referred to as: B4, B3A, B2A2, BA3 and A4. They are also referred to as LDH1 to LDH5 respectively based on electrophoretic mobility from anode to cathode.

H4/B4 is predominantly in heart -an aerobic tissue- this isoenzyme has low K_m i.e. high affinity for pyruvate and is allosterically inhibited by high levels of pyruvate. It favours the reaction from lactate to pyruvate. M4/A4 tetramer is primarily present in tissues that can experience anaerobic conditions e.g. Skeletal muscle. This isoenzyme has a higher K_m i.e. low affinity for pyruvate and is not inhibited by it. It favours the reaction from pyruvate to lactate. Other tissues have a mixture of all the five isoenzymes and have kinetic properties depending on the ratio of the different sub-units in the tetramer. In serum all the five isoenzymes can be detected.

3

GLUCONEOGENESIS

General Introduction

It refers to the process of synthesising glucose from non-carbohydrate precursors like lactate, glycerol, oxaloacetate, propionate etc. and alanine or from other selective group of amino acids called gluconeogenic amino acids. This process takes place in three different compartments *viz.* cytosol, mitochondria and endoplasmic reticulum. As mentioned above glycolysis and gluconeogenesis should not operate simultaneously as it constitutes a futile cycle. Hence they are reciprocally regulated. That is the enzymes which are active during glycolysis must be inhibited when gluconeogenesis is operative. Gluconeogenesis cannot simply take place by the exact reversal of glycolytic reactions. As mentioned above there are three irreversible steps in glycolysis. These steps have to be overcome and they are carried out by a different set of reactions unique to gluconeogenesis. Gluconeogenesis takes place when levels of ATP are high.

Gluconeogenic reactions - Bypassing the irreversible steps in glycolysis

(a) PK reaction is reversed and Phosphoenolpyruvate (PEP) formed in the cytosol by the help of four reactions (3 different enzymes). (a) First pyruvate arising from any source enters mitochondria (in case of eukaryotes) and (i) pyruvate carboxylase converts it to oxaloacetate:

Pyruvate + CO_2 +ATP $\rightarrow$ oxaloacetate + ADP.

(ii) In the second step Malate dehydrogenase (MDH) in mitochondria reduces oxaloacetate to malate:

Oxaloacetate + NADH + H^+ ↔ Malate + NAD^+; Then malate is transported out of mitochondria into cytosol and reconverted to oxaloacetate by the reversal of the above reaction by cytosolic MDH.

In the last step (iii) oxaloacetate is decarboxylated at the expense of an energy rich compound GTP and forms PEP by the cytosolic enzyme PEP carboxy kinase; oxaloacetate + GTP → CO_2 + GDP + PEP. This enzyme can also be present in mitochondria depending on the organism and hence PEP can be formed right in mitochondria and transported into cytosol. By these reactions the first reversible step in the reverse direction (last step in the direction of glucose breakdown) is overcome.

(b) Next step to be circumvented is that of PFK-1. This is accomplished by F1,6, BPase which hydrolyses F1,6, BP into F-6-P; F1,6, BP +H_2O → F-6-P + Pi. This activity and PFK-1 activity reside on the same polypeptide. But the reverse reaction is facilitated by inactivating the PFK-1 and activating F1,6, BPase. In *E.coli* mutants are available which are deficient in F1,6, BPase. They cannot grow on any other carbon sources. This indicates the importance of F1,6, BPase for gluconeogenesis.

(c) The first irreversible step of glycolysis is bypassed by the action of Glucose -6-Phosphatase:

G6P + H_2O → Glucose + Pi. This enzyme is bound to endoplasmic reticulum in eukaryotes. Hence G6P enters the lumen of endoplasmic reticulum where it is hydrolysed. The products are then transferred to cytosol.

So in the above manner the three irreversible steps of glycolysis are overcome and the cell is able to synthesise glucose.

Regulatory details of gluconeogenic enzymes

Now as mentioned before both breakdown and synthesis of glucose should not occur simultaneously to avoid futile cycle. So the enzymes of glycolysis and gluconeogenesis must not be active at the same time and they should be reciprocally regulated. In fact, it is accomplished very economically by using at different places the same metabolites. Thus, F2,6 BP and AMP which are activators of PFK-1 allosterically inhibit F1,6, BPase. Hence both enzymes cannot be active at the same time. Further gluconeogenesis takes place when levels of ATP are high which inhibit PFK-1. Glucagon hormone by a signal transduction cAMP cascade simultaneously phosphorylates both PFK-2 and FBPase-2. Phosphorylation makes the PFK-2 inactive while the same modification i.e. phosphorylation makes the FBPase 2 active. Therefore, glucagon decreases F2,6, BP levels and inhibits glycolysis and stimulates gluconeogenesis. While dephosphorylation activates PFK-2, dephosphorylation inhibits FBPase-2. Therefore, it results in the increase in the concentration of F 2,6, BP and stimulation of glycolysis and inhibition of gluconeogenesis. Insulin acts in an opposite manner. It activates phosphoprotein phosphatase that catalyzes the removal of the phosphoryl groups simultaneously form both PFK-2 and FBPase-2. As mentioned above both enzymes form a bifunctional polypeptide. Phosphorylation has opposing effects on PFK-2 and FBPase-2. As a result of phosphorylation PFK-2 activity is increased and FBPase-2 activity is decreased, leading to increased levels of the metabolite F2,6 BP causing stimulation of glycolysis and inhibition of gluconeogenesis. Epinephrine also increases levels of F2,6 BP which is a strong activator of PFK-1 thereby increasing glycolysis. Enzymes which phosphorylate proteins are called Protein Kinases. Enzymes which remove phosphate groups from proteins are called as Protein Phosphatases or Phosphoprotein phosphatases.

Pyruvate kinase inactivation by phosphorylation induced by hormones glucagon and vasopressin inhibits glycolysis and stimulates gluconeogenesis. Pyruvate carboxylase is allosterically activated by acetyl CoA in mitochondria. Acetyl CoA levels increase because of increased lipolysis followed by beta-oxidation of fatty acids which give rise to acetyl CoA. Indirectly hormonal control is involved as epinephrine promotes lipolysis and beta-oxidation and insulin inhibits lipolysis. PEP carboxykinase, Glucose 6-phosphatase, PFK-1, PK, PFK-2/FBPase-2 are also controlled at transcriptional level by the above mentioned hormones. High levels of $NADH+H^+$ inhibits gluconeogenesis as $NADH+H^+$ can be used for the synthesis of lactate from pyruvate or malate form oxaloacetate. Hence pyruvate and oxaloacetate are not available for gluconeogenic enzymes.

4

TRICARBOXYLIC ACID/CITRIC ACID/ KREBS CYCLE (REFERRED IN THE FOLLOWING DISCUSSION AS TCA CYCLE)

General Introduction

In Glycolysis glucose, strictly speaking is not oxidized. If Glucose ($C_6H_{12}O_6$) is metabolised to two molecules of lactate i.e. 2X CH_3–CHOH–COOH equals to $C_6H_{12}O_6$ i.e. glucose. So what happened is an intramolecular oxidation. And we get energy gain only to the extent of 2 ATP. But glycolysis can supply energy rapidly in times of special needs. However, the organism need to harvest more energy from glucose by its complete oxidation. For this, in the presence of oxygen the two pyruvate (CH_3–C=O–COOH) molecules enter mitochondria and get oxidized to give much more energy required for the organism. Further the oxidation of pyruvate through a series of reactions also produces intermediates that can be used in the synthesis of several other required metabolites. Hence TCA cycle is not only considerably energy yielding but also serves as an amphibolic pathway. Pyruvate thus links glycolysis to TCA cycle. TCA cycle is not only a pathway for the oxidation of carbohydrates but also for fatty acids and amino acids. Thus, it is a common pathway for carbohydrates, fatty acids (lipids) and amino acids (proteins). All the three categories of molecules can be used for the generation of energy via TCA cycle and in the process can also generate several useful intermediate metabolites which act as precursors for other compounds. Pyruvate after entering into mitochondria is oxidatively decarboxylated to form acetyl CoA, a reaction catalyzed by Pyruvate Dehydrogenase (PDH) Multienzyme complex. Multienzyme complex may be distinguished from a multifunctional polypeptide where the different polypeptides are covalently linked together into a large single polypeptide. In multienzyme complex many enzymes are just associated with each other and they can be separated. The overall reaction is a result of different sub-reactions. PDH consists of PDH (E1) which catalyzes decarboxylation of pyruvate, dihydrolipoyl transacetylase (E2) which transfers the acetyl group to CoA and dihydrolipoyl dehydrogenase(E3) which catalyzes the regeneration of the oxidized form of lipoic acid. The three enzymes require TPP, lipoic acid and FAD respectively. TPP and FAD are bound to E1 and E3 respectively. CoA and NAD are substrates for E2 and E3 respectively. Lipoic acid is covalently bound to a lysine reside on E2-hence referred to as lipoamide. All cofactors are regenerated to their original form at the end of the overall reaction. The complex also contains two enzymes that regulate the enzyme activity *viz.*PDH kinase and PDH phosphatase particularly in eukaryotes. In *E.coli*, PDH has a molecular weight of 5 million daltons consisting of 60 polypeptide chains. But apparently there is no covalent regulation by phosphorylation. Molecular weight of PDH multienzyme complex may vary from 4 to 10 million daltons among organisms. Like PDH, alpha KGDH also is a large multienzyme complex.

Plants have PDH complex both in mitochondria and in plastids and like in mammalian systems is inhibited by the products, NADH and acetyl CoA. Mitochondrial enzyme is also regulated by reversible covalent modification by phosphorylation.

Overall reaction catalyzed by the PDH multienzyme complex is:

Pyruvate + CoA + NAD $\rightarrow$ acetyl- CoA + NADH + CO_2. Enzymes associated with the complex catalyzes sub-steps finally giving the above overall reaction.

(Students may be clear about the nomenclature: in ionised form the terms pyruvate, lactate, oxaloacetate etc are used. In unionised forms they are referred to as pyruvic acid, lactic acid and oxaloacetic acid respectively).

Acety-CoA released by PDH complex reacts with oxaloacetate to give citrate-a tricarboxylic acid. It is called a cycle as the oxaloacetate consumed in this reaction is regenerated after a series of reactions and the regenerated oxaloacetate once again combines with another molecule of acetyl-CoA and completes another round of reactions and so on.

Regulatory details of different enzymes

TCA cycle is regulated mainly at the steps of PDH, Citrate synthase, Isocitrate dehydrogenase (ICDH) α-ketoglutarate dehydrogenase (α KGDH -also called 2-oxoglutarate dehydrogenase). The enzymes are regulated allosterically, by covalent modification, energy status and catabolite repression.

(a) PDH

PDH is both allosterically and covalently regulated. The enzyme is allosterically activated by NAD, AMP and CoA and Ca^{2+} and inhibited by acetyl CoA, NADH (product inhibition) and ATP. In other words, indicators of high energy availability, inhibit the reaction. The PDH kinase a part of the multienzyme complex can phosphorylate PDH thereby inactivating it. PDH kinase phosphorylates a serine residue near to the C-terminal end. Multienzyme complex also contains PDH phosphatase which can remove the phosphoryl group and the inhibition is reversed. These two are categorised as regulatory enzymes. The regulatory enzymes themselves are subjected to regulation. PDH kinase is activated by acetyl CoA, NADH, and inhibited by pyruvate, ADP and high magnesium (Mg^{2+}) ions and potassium (K^{+}) ions. PDH phosphatase activated by magnesium (Mg^{2+}) ions and calcium (Ca^{2+}) ions. In general, it may be said that high ratios of ATP/ADP, NADH/NAD and aetyl- CoA/CoA leads to phosphorylation and subsequent inactivation of PDH complex.

(b) Citrate synthase

This catalyzes the reaction -acetyl-CoA + oxaloacetate + $H_2O \rightarrow$ citrate + CoA. The enzyme is allosterically inhibited by ATP, NADH, and acetyl CoA -all indicate a high energy status. Hence no need for generating further energy. Accordingly, it is activated by NAD and

CoA. Further citrate and succinyl CoA are competitive inhibitors for oxaloacetate and acetyl CoA.

(c) ICDH

It catalyzes the oxidative decarboxylation of isocitrate to α-ketoglutarate (also called 2-oxoglutarate) with the intermediate formation of oxalosuccinic acid. So reaction can be written as:

Isocitrate + $NAD^+ \rightarrow$ NADH + H^+ (oxalosuccinate, temporary intermediate) $\rightarrow$ alpha-ketoglutarate + CO_2

ICDH is allosterically activated by ADP and inhibited by ATP and NADH. ICDH is also inactivated by phosphorylation both in pro and eukaryotes. Dephosphorylation by phosphatase reactivates it. In *E.coli* it is phosphorylated at serine 113 (numbering done from N-terminal amino acid as number one). It is also activated by Ca^{2+}.

(d) α-KGDH

It catalyzes the reaction; α-ketoglutarate + CoA + $NAD^+ \rightarrow$ Succinyl-CoA + CO_2 + NADH + H^+.

The enzyme like the previously mentioned is inhibited by high energy charge i.e. ATP and NADH. Also inhibited by Succinyl CoA called product inhibition. Enzyme activated by AMP which indicates a low energy status.

(e) Other regulatory mechanisms

TCA cycle enzymes may exist as metabolons facilitating the transfer of product of one reaction to the next enzyme and so on. In *E.coli* many citric acid cycle enzymes are regulated by catabolite repression.

Energy production—Energy production in the TCA cycle occurs in several forms

(a) GTP- Succinyl CoA synthetase catalyzes the reaction

Succinyl- CoA +GDP $\rightarrow$ Succinate + GTP + CoA. GTP is an energy rich compound and is equivalent to ATP. If need be it can be converted to ATP by reacting with ADP and in the process GTP becomes GDP. This reaction as mentioned before is catalyzed by NDP kinase.

Succinyl CoA synthetase reaction is an example of substrate level phosphorylation occurring in mitochondria where oxidative phosphorylation naturally occurs forming only ATP.

(b) ATP - succinyl CoA synthetase

In some animals a different isoenzyme of succinyl CoA synthetase uses ADP instead of GDP and therefore forms ATP. So is with yeast.

(c) NADH

When NADH passes through the mitochondrial electron transport and is oxidized we get ATP at specific sites in the electron transport chain and H_2O at the end.

(d) $FADH_2$

Like NADH it also gives ATP and water. Here students must be clear about the number of ATP's formed when NADH and $FADH_2$ are passed through electron transport chain. For a very long time the number of ATP for NADH oxidation and $FADH_2$ oxidation via electron transport chain was thought to be 3 (three) and 2 (two) respectively. Even today some books mention so. But the right figure is **2.5 ATP** and **1.5 ATP** for NADH and $FADH_2$ oxidation. Hence energy calculations must be made accordingly. NADH and $FADH_2$ are also referred to as reducing power as they can be used in metabolic reactions as a source of reductant.

Amphibolic nature and interrelationships with biosynthetic pathways of other metabolites

(a) General Introduction

As mentioned above TCA cycle generates several useful intermediates which can be used as precursors for other compounds and hence TCA cycle is interconnected with other metabolic pathways. Hence, TCA cycle intermediates are removed/depleted from the cycle as and when and to the extent required. Hence, TCA cycle is regarded as an amphibolic pathway. This removal of intermediates is termed as cataplerotic reactions. Main intermediates used as biosynthetic precursors are: citric acid for the biosynthesis of fatty acids and cholesterol; Succinyl CoA for the biosynthesis of porphyrins;

α-ketoglutarate and oxaloacetate for the biosynthesis of glutamic acid and aspartic acid respectively. The above products formed from TCA cycle intermediates can in turn be used for the biosynthesis of other molecules. Malate also can be removed for the subsequent synthesis of glucose.

Oxaloacetate by MDH; Malate $+NAD^+ \rightarrow$ Oxaloacetate + NADH + H^+. Alternatively, malate can also be oxidatively decarboxylated to form pyruvate. The converse of cataplerotic reactions is anaplerotic reactions also called as refilling or replenishing reactions. The intermediates that are removed from either glycolysis or TCA cycle can be replenished by certain other set of reactions/pathways.

(b) Anaplerotic reactions—Main anaplerotic reactions/pathways are; amino acids isoleucine, methionine and threonine, valine and odd chain fatty acids can be metabolised to give succinyl CoA; amino acids aspartate, phenylalanine and tyrosine can give rise to fumarate; Asparagine and aspartate can provide oxaloacetate; amino acids arginine, glutamine, glutamic acid, histidine and proline can be degraded and give rise to α-ketoglutarate. Pyruvate by carboxylation can give rise to oxaloacetate a reaction catalyzed by pyruvate carboxylase. Oxaloacetate can also be produced by the action of PEP carboxylase;

PEP + $HCO^{3-} \rightarrow$ oxaloacetate + Pi. As mentioned above under gluconeogenesis PEP the glycolytic intermediate can be generated from oxaloacetate by the PEP carboxykinase reaction which is reversible. Oxaloacetate can also be formed by MDH:

Malate +NAD $\rightarrow$ Oxaloacetate + NADH, a reversible reaction. Alternatively, malate can also be oxidatively decarboxylated to form pyruvate; Malate + $NADP^+ \rightarrow$ Pyruvate + CO_2 + NADPH + H^+. This is also a reversible reaction catalyzed by malic enzyme (Don't confuse between malic enzyme and malate dehydrogenase). Involvement of the above enzymes forming/utilizing oxaloacetate may vary among organisms. PEP carboxylase and malic enzyme are very common in plants. As mentioned under glycolysis, it is interconnected with oxidative pentose phosphate (OPP) pathway. The intermediates F-6-P and GA3P generated in OPP pathway can enter glycolytic pathway. Alanine, cysteine, glycine, serine, threonine and tryptophan on degradation can give pyruvate, the glycolytic metabolite.

Based on the above anaplerotic reactions amino acids may be classified as gluconeogenic or ketogenic or both gluconeogenic and ketogenic depending on their degradative pathways. All those amino acids which can give rise to glucose has to end up in a glycolytic or TCA cycle intermediate and thus serve an anaplerotic role. Glycine, serine, valine, histidine, arginine, cysteine, proline, alanine, glutamic acid, glutamine, aspartate, asparagine and methionine are exclusively gluconeogenic. Phenyl alanine, isoleucine, threonine, tyrosine and tryptophan are both gluconeogenic and ketogenic. Leucine and lysine are exlcusively ketogenic i.e. they give rise to acetoacetyl CoA and acetyl CoA which results in the biosynthesis of ketone bodies *viz.* acetoacetate, betahydroxybutyrate and acetone.

(c) Cataplerotic reactions

Amino acids may also be classified in terms of their biosynthetic families on the basis of the carbon and/or amino acid precursor.

(a) *Serine family*- The glycolytic intermediate 3-Phosphoglyeric acid (PGA) is used for the biosynthesis of serine which in turn can be used for the synthesis of glycine and cysteine.

(b) *Glutamate family*- The TCA cycle intermediate α-ketoglutarate initially gives rise to glutamic acid which in turn can be used for the synthesis of glutamine, proline, and arginine.

(c) *Aspartate family*- The TCA cycle intermediate oxaloacetate (also can be formed in cytosol and chloroplast) initially forms aspartate that can be used for the synthesis of asparagine, lysine, threonine, and isoleucine

(d) *PEP family*-PEP is a glycolytic intermediate which can act as a precursor for the synthesis of aromatic amino acids, phenyl alanine, tyrosine and tryptophan.

(e) *Pyruvate family*- Pyruvate is the glycolytic metabolite and can be used for the synthesis of alanine, valine and leucine.

(f) Histidine is derived from ribose-5-phosphate formed in OPP pathway.

Amino acids formed as above are in turn used for synthesis of all proteins, so many other nitrogenous metabolites, and also as precursors for purine and pyrimidine synthesis and thereby RNA and DNA synthesis. By this statement one can appreciate the statement that metabolism is highly interconnected. Hence regulation of one pathway has an influence on several other metabolic pathways.

Students should be clear that though only 20 amino acids have been mentioned above, two more amino acids have been added which can be incorporated into proteins in response to a genetic code. They are seleno cysteine and Pyrrolysine. Hence we have 22 amino acids which can be incorporated into proteins in response to genetic code (a codon in mRNA). Apart from the above we have hundreds of non-protein amino acids particularly found in plants. In animal systems ornithine and citrulline are very common.

5

Glyoxylate Cycle and its Regulation

Plants and some bacteria have the capacity to synthesise glucose from fatty acids. Animal systems do not possess this capacity though they can accumulate large amounts of triglycerides. Triglycerides after hydrolysis liberates three molecules of fatty acids and one molecule of glycerol. The glycerol portion (which constitutes only about five per cent of the triglyceride portion) can be converted in to glucose by all organisms including animal systems. But not acetyl CoA. This is because the carbon atoms of acetyl CoA fragment resulting from β-oxidation of fatty acids are ultimately lost when metabolised by TCA cycle. Alternatively, acetyl CoA can be used to form ketone bodies which can be used as a fuel by brain which cannot use any other metabolic fuel other than glucose. During starvation ketone bodies contributes to the energy requirements of the brain. However, the carbons can be conserved and used for glucose synthesis if glyoxylate cycle operates which bypasses two TCA cycle enzymes which leads to loss of two CO_2. Hence TCA cycle and glyoxylate cycle are coordinately regulated. The cycle in case of plants takes place in special sub-cellular organelles called glyoxysomes which also carry out β-oxidation of fatty acids. The cycle is of importance during seed germination, particularly oil seeds when the stored triglycerides are converted to glucose for further carbohydrate metabolism. The enzymes unique to glyoxylate cycle are isocitrate lyase (ICL) and malate synthase. In all, four compartments are involved in the entire process -lipid bodies, glyoxysomes, mitochondria and cytosol. In prokaryotes all reactions take place in the same compartment.

In lipid bodies triglycerides are hydrolyzed to free fatty acids and glycerol. Fatty acids then enter glyoxysomes undergo β-oxidation forming acetyl CoA and glyoxylate cycle. However, glyoxysomes lack aconitase enzyme. So after citrate formation, citrate is likely to be converted to cis-aconitate and isocitrate in cytosol and isoitrate transported back to glyoxysomes. (Aconitase is present both in cytosol and mitochondria). Bacteria and plants can also synthesize acetyl CoA from free acetate by acetyl CoA synthetase: acetate + CoA + ATP → acetyl CoA + AMP + PPi (inorganic pyrophosphate). PPi is subsequently hydrolysed to 2Pi by inorganic pyrophosphatase. (Students should remember that in metabolism in many reactions ATP is cleaved into AMP and PPi. Such reactions can be pulled forward i.e. go to completion only after the subsequent breakdown of PPi by inorganic pyrophosphatase. Further when ATP is cleaved as AMP and PPi you have to take into calculations that two ATP's are consumed in the reaction). Succinate formed in glyoxysomes enters mitochondria and is converted to malate. This malate is exported to cytosol for gluconeogenesis by the reactions mentioned earlier. In the conventional TCA cycle acetyl CoA combines with oxaloacetate and forms citrate which is further converted to isocitrate. Isocitrate so formed is oxidatively decarboxylated by ICDH to form alpha-ketoglutarate and a carbon is lost as CO_2. Glyoxylate cycle bypasses this reaction. Instead the organism cleaves isocitrate into glyoxylate and succinate by the enzyme isocitrate lyase. Glyoxylate combines with another molecule of acetyl CoA and water and forms malate by the action of the enzyme malate synthase. Malate so formed can be oxidised to oxaloaetate by MDH and combine with another molecule of citrate to start the cycle again. Succinate formed as above enters mitochondria and converted by TCA cycle reactions to malate via fumarate i.e. by the successive actions of succinate dehydrogenase (SDH) and Fumarase as follows:

Succinate + FAD → Fumarate + $FADH_2$ followed by Fumarate + H_2O → Malate. Malate so formed is exported into cytosol and enters the gluconeogenic pathway as described above.

Overall reaction for glyoxylate cycle is represented as 2(acetyl CoA) + NAD^+ + $2H_2O$ $\rightarrow$ succinate + 2 CoA + NADH+ H^+.

ICDH and ICL are coordinately regulated. When ICDH is phosphorylated by protein kinase it becomes inactive and therefore is isocitrate cannot be oxidatively decarboxylated. Then ICL acts on isocitrate giving glyoxylate and succinate. On dephosphorylation by protein phosphatase ICDH it is reactivated and isocitrate is not available for ICL but metabolised by ICDH followed by α-KGDH resulting in the loss of two carbons in the form of CO_2. If ICL is active neither isocitrate is decarboxylated nor α KG is formed which can be decarboxylated by α-KGDH. So no carbons are lost. ICDH is inhibited by AMP and ADP while they activate ICL. Glycolytic and citric acid cycle intermediates, AMP, ADP, PEP, pyruvate, 3 PGA, isocitrate, α-KG and oxaloacetate stimulate phosphatase activity.

6

Oxidative Pentose Phosphate (OPP) Pathway

General Introduction and functions

Often it is simply called as PP pathway. But to distinguish it from Calvin cycle (photosynthesis) which is referred to as reductive PP pathway in chloroplasts it is called OPP pathway which takes place in cytosol. This pathway is also referred to as HMP (hexose monophosphate) pathway or shunt. Glucose can also be oxidized by this pathway. In every round one CO_2 is lost. So in six rounds, six CO_2 can be lost, equivalent to complete oxidation of glucose. However, the role of this pathway is not to produce energy but, (A) generate reducing power in the form of NADPH in contrast to NADH and $FADH_2$ in TCA cycle. This NADPH is used as a reductant in several metabolic reactions/pathways for example fatty acid biosynthesis, cholesterol biosynthesis, (B) generate pentoses to provide ribose and deoxy ribose for RNA and DNA, nucleotides, CoA etc., (C) provide the precursor Ribose 5-phosphate for the biosynthesis of histidine, (D) provide erythrose 4-phosphate which along with glycolytic intermediate PEP is used for the biosynthesis of aromatic amino acids, (E) to facilitate interconversion of 3, 4, 5, 6 and 7 carbon sugars, (F) if need be intermediates F6P and GA3P can be shuttled into glycolysis. Hence it is clear from "A" to "F" that OPP pathway is interconnected to so many other pathways and hence its regulation can influence so many other pathways. The overall reaction of OPP pathway is represented as follows:

G6P + 2NADP$^+$ +H_2O → ribose-5-phosphate + 2 NADPH + 2H$^+$ + CO_2.

Pathway reactions and Regulatory details

(a) Glucose-6-phosphate dehydrogenase (G6PDH)

G6PDH is the rate limiting step and is the major control point. Along with a lactonase it catalyzes the essentially irreversible reaction:

G6-P +NADP +H_2O → 6-phosphogluconate + NADPH + H$^+$ with the intermediate formation of 6-phosphoglucono-Δ-lactone. After the hydrolysis of lactone by lactonase the reaction is complete and we get 6-phosphogluconate. The enzyme also requires magnesium or manganese or calcium divalent ions. The enzyme is stimulated by NADP and strongly product inhibited by NADPH. It is also inhibited by fatty acyl CoA. As Insulin hormone stimulates many anabolic reactions requiring the reducing power NADPH, it stimulates G6PDH.

(b) Rate of OPP pathway

Rate of OPP pathway depends on the need for reducing power NADPH, ribose-5-phosphate (R5P) and ATP. In rapidly dividing cells there is more need for R5P for the synthesis of nucleic acids. But during lactation more amount of reducing power NADPH is required for the synthesis of fatty acids in the mammary gland. In those organisms where adipose is not there the synthesized fatty acids may be used to become component of membranes. Broadly we can think of four different situations: (a) when more R5P than NADPH is required. (b) when both R5P and NADPH are sufficiently required (c) when more NADPH than R5P is required and (d) when both NADPH and ATP are required. Accordingly, OPP pathway is regulated to meet the above situations. To understand the regulation in the above situations it is necessary to know the entire set of reactions of OPP pathway. After the first two reactions mentioned above by G6PDH and lactonase resulting in the formation of 6-phosphogluconate, subsequent reactions are as follows: Above, two reactions of the pathway are mentioned. The third reaction is:

6-phosphogluconate + $NADP^+ \rightarrow$ Ribulose-5-phosphate + CO_2 + NADPH +H^+ catalyzed by 6-phosphogluconate dehydrogenase (6PGDH). So in this reaction the first pentose phosphate is formed in the form of a ketose. Six carbon glucose has become a five carbon compound because of loss of a carbon in the form of CO_2. In the fourth and fifth reactions Ribulose -5-phosphate (Ru5P) is isomerised to R5P and epimerised to Xylulose-5-phosphate by an isomerase and epimerase respectively.(of course both enzymes broadly come under the category of isomerases the sixth group in the enzyme classification).

The seventh and eighth reactions involve the enzymes Transketolase and Transaldolase.

Ru5P + Xylulose-5-Phosphate $\leftrightarrow$ Sedoheptulose 7-phosphate + GA3P (Reaction catalysed by Transketolase (TK); Sedoheptulose 7-phosphate + GA3P $\leftrightarrow$ Erythrose-4-Phosphate + F6P (Reaction catalyzed by Transaldolase (TA). Ninth reaction is again catalyzed by TK as follows:

Erythrose-4-Phosphate + Xylulose-5-Phosphate $\leftrightarrow$ F6P + GA3P.

Transketolase (TK) is an enzyme which catalyzes the transfer of two carbons from a ketose to an aldose. In the process the former ketose is shortened by two carbons and becomes an aldose while the former aldose is lengthened by two carbons and becomes a ketose. TK catalyzes two different reactions in OPP pathway as shown above.

Both TK and TA reactions are reversible reactions and hence students should accordingly visualize the reactions.

Transaldolase (TA) is an enzyme which transfers three carbons from a ketose to an aldose. The result is as in the case of TK but here the lengthening and shortening of the metabolite is by three carbons. Hence, TK and TA are involved in aldose to ketose interconversion and reversibly lengthening or shortening the resulting metabolites.

Salient points to be noted from the above nine reactions are: (i) the first two reactions are essentially irreversible, (ii) The remaining seven reactions are freely reversible, (iii) the first two reactions are referred to as oxidative phase of OPP pathway and it is in these reactions NADPH $+H^+$ is formed apart from evolution of CO_2, (iv) the remaining seven reactions are referred to as non-oxidative phase of OPP pathway. It is in these reactions R5P required for nucleotides, and glycolytic intermediates F6P and GA3P are formed. Glycolytic intermediates formed can enter into glycolytic pathway. From the above it is broadly evident how OPP pathway may be regulated. When the need for NADPH $+H^+$ is more the oxidative phase is more active and when the need is for R5P is more the non-oxidative phase is more active. When both reducing power and R5P are required both phases have to be active. When there is a need for ATP and no much need for reducing power or R5P the glycolytic intermediates F6P and GA3P are channeled towards glycolysis.

In the first situation when more R5P is required TK and TA preferably operate in the direction of formation of R5P. In the second situation when both R5P and NADPH are required, the first four reactions of OPP pathway predominates. The third situation demands more NADPH than R5P. Here the oxidative reactions predominate. In all the first four reactions play an important role. In the fourth situation when both NADPH and ATP are required the oxidative phase is active generating NADPH $+H^+$ and the intermediates F6P and GA3P formed by TK/TA enter into glycolysis to increase the supply of ATP.

7

Regulation of Glycogen Degradation and Biosynthesis

General introduction

Glycogen is a reserve carbohydrate in animals similar to starch in plants. (However, fungi also may have glycogen). It is primarily stored in liver and muscle and to a little extent in other tissues. Glycogen has an amylopectin structure i.e. is a linear chain of glucose units linked alpha $1 \rightarrow 4$ with branches of alpha $1 \rightarrow 6$ linkages. Compared to starch the degree of branching is more in glycogen - one in 8 to 12 residues while in starch it can vary from 20 to 30 residues. The length of the branches also is shorter in glycogen. Though as percentage glycogen content is more in liver -5 to 10 per cent and in muscle 1 to 2 per cent, the total amount of glycogen in muscles is much more than in liver as the total mass of the muscle is more compared to liver mass. The purpose of storage is different in muscle and liver. In muscle the primary purpose is to provide energy from glycogen. Whereas in liver the purpose is to supply free glucose to blood and brain. However, in both the tissues both glycogen degradation and synthesis take place depending on the metabolic condition. Both these processes are reciprocally regulated. Before we try to understand the regulation, an understanding of metabolic pathways of glycogen synthesis and break down would be useful.

Glycogen biosynthesis

One of the striking differences between glycogen synthesis and starch synthesis is the nature of primer required. While in starch an oligosaccharide may serve as a primer i.e. for further addition of glucose units and lengthening of the chain, in glycogen synthesis the primer is a protein (37kDa) called glycogenin. Glycogenin is catalytically active and can catalyze the auto addition of glucose units to the hydroxy group of tyrosine 194 in the glycogenin protein. The first glucose is attached to hydroxy group by carbon 1 of glucose. Glycogenin can add to itself up to 7 or more residues. Such a glycogenin protein with an oligosaccharide acts as for glycogen synthase making a linear chain of glucose units of alpha 1 → 4 linkages. The donor of glucose units is UDPG (uridine diphosphate glucose) catalyzed by UDPG pyrophosphorylase. By the combined action of the above enzymes a long linear chain of varying length is formed. Reactions may be represents as:

UTP + G1P → UDPG + PPi (as mentioned elsewhere PPi is hydrolysed to two Pi by pyrophosphatases which help the reaction to proceed to completion i.e. pulls the reaction in the forward direction).

Glycogenin primer (formed as above) + UDPG → glycogen primer plus one more residue of glucose + UDP. The addition of glucose units in these manner continues by the action of glycogen synthase and the chain formed is of varying lengths.

UTP is regenerated in the NDP kinase reaction - UDP + ATP → UTP + ADP

Branches i.e. glucose units linked to the main chain by alpha 1 → 6 linkages are created by the branching enzyme by transferring glucose units from the main chain to a specific glucose unit on it.

Glycogen degradation

Both in liver and muscle glycogen is broken down by phosphorolytic cleavage of alpha 1 → 4 linkages from the main chain, to from G1P and a glycogen molecule shorter by one unit in the main chain by glycogen phosphorylase. This reaction is repeated to give more and more of required glucose units.

Glycogen + Pi → G1P + Glycogen shorter by one glucose unit.

G1P formed as above can have different fates. In liver it is isomerised to G6P by phosphoglucomutase and subsequently hydrolyzed by glucose-6-phosphatase to form free glucose and Pi. The Pi is recycled in the phosphorylase reaction. Free glucose liberated is exported to blood to raise the glucose concentration from where it can reach other tissues. Muscles do not possess glucose-6-phosphatase and hence G6P is metabolised *via* glycolysis and as mentioned above the purpose of glycogen breakdown in muscle is to provide energy when it is exercised. The catalytically active form of the enzyme is a dimer (alpha-beta).

Branches are degraded by the help of three enzymes. (i) Phosphorylase can remove glucose units as G1P till about four glucose units are remaining in the branch, (ii) Then another enzyme glucan transferase transfers a trisaccharide unit from the branch to the main chain or another oligosaccharide - a transglycolase reaction. The links formed by this enzyme are alpha 1→ 4 between glucose units. Phosphorylase cannot act on the terminal glucose on the branch having alpha 1→ 6 linkages, (iii) The debranching enzyme then removes the terminal glucose with alpha 1→ 6 linkages - not phosphorolytically using Pi but hydrolytically using a molecule of water - and forms free glucose. About 90 per cent of the glucose units are formed as G1P and the remaining 10 per cent as free glucose. This free glucose in case of liver can directly be exported to blood while in muscle it needs to be phosphorylated to enter glycolysis.

Regulation

Glycogen degradation and biosynthesis are reciprocally regulated i.e. both processes are not active at the same time. That would result in a futile cycle wasting energy of UTP. (i) Phosphorylase in both liver and muscle are covalently regulated by phosphorylation. The enzyme exists in two forms -active phosphorylase termed phosphorylase-a and inactive or very less active form termed phosphorylase-b. An enzyme phosphorylase kinase phosphorylates serine 14 in each subunit of the dimer and converts the inactive phosphorylase-b into active phosphorylase-a. The enzyme phosphorylase kinase itself exists in both active and inactive forms a and b. Glycogen synthase exists in two forms active and inactive (or much less active) form. The inactive form is covalently phosphorylated at serine residues. It is not one but as many as 9 serine residues can be phosphorylated and this phosphorylation can be carried out by four different protein kinases. With every successive phosphorylation the enzyme becomes more and more inactive. This form is also termed Glycogen synthase-b or Glycogen synthase D ("D" indicating dependency on Glucose 6- phosphate for its activity). This form is activated both by Glucose 6- phosphate and ATP. The other form of Glycogen synthase is active and is unphosphorylated. Unlike Phosphorylase, Glycogen synthase becomes active as a result of dephosphorylation. Dephosphorylation is carried out by protein phosphatase -1(PP-1). This form is also termed Glycogen synthase a or Glycogen synthase I ("I" indicating that it is independent of Glucose 6- phosphate for activation). When energy level -ATP levels -is high -glycogen synthesis is favoured.

Similarly, high levels of glucose 6-phosphate which indicates high levels of glucose activates glycogen synthase and glycogen synthesis. High levels of ATP and Glucose 6-phopshate allosterically inhibits phosphorylase, but AMP activates. The above phosphorylations and dephosphorylations are mediated by hormones. Epinephrine and glucagon via cAMP cascade (see signal transduction) and protein kinase A phosphorylates the enzyme phosphorylase and activates it while glycogen synthase is inactivated by phosphorylation. As mentioned above glycogen synthase is phosphorylated by other protein kinases too. Insulin by activating phosphodiesterase removes the phosphate groups thereby inactivating phosphorylase and activating glycogen synthase. One important regulatory differences between glycogen phosphorylase in muscle and liver is that glucose inhibits the liver enzyme but not the muscle enzyme, and Ca^{2+} stimulates the muscle enzyme but not the liver enzyme.The primary purpose of glycogen breakdown in muscle is to ensure rapid supply of energy when needed whereas in liver the primary purpose of glycogen breakdown is to ensure supply of free glucose to other tissues when needed.

Part - III

Regulation of Lipid Metabolism

8

Triacylglycerol and Fatty Acid Break down

General Introduction

For our discussion we are referring to Lipid in the sense of Triglycerides or also called by relatively recent nomenclature Triacylglycerols. A triacylglycerol is a molecule of glycerol esterified to three fatty acids separately at the three carbon atoms of glycerol. It is the hydroxy group of glycerol which participates in the esterification reaction. The three fatty acids can be identical or different. Not necessary all three are different but two can be identical and one can be different. If all three are identical accordingly they can be called as stearin or tristearin derived from stearic acid, palmitin or tripalmitin derived from palmitic acid etc.

Fatty acids can be saturated or unsaturated and can be of varying length depending on the source. The most common fatty acids are of saturated 16 (palmitic acid) or 18 carbon (stearic acid) length, but longer fatty acids particularly unsaturated are very important physiologically. USFA of 18 carbon length serve as EFA. USFA of 16 carbon also are found (palmitoleic acid). The double bonds in the naturally occurring fatty acids are of Cis-configuration. Animals can form only one unsaturated fatty acid (USFA) - oleic acid, a mono unsaturated fatty acid with double bond between carbon 9 and carbon 10. But animals cannot synthesize poly unsaturated fatty acids but are essential for them and hence have to be supplied in diet. These are called essential fatty acids (EFA) or polyunsaturated fatty acids (PUFA) *viz.* linoleic and linolenic acids with two (at carbons 9 and 12) and three double bonds (at carbons 9,12 and 15) respectively. They are also called omega-6 and omega-3 fatty acids the omega carbon referring to the last carbon of the methyl group. The general structure of fatty acid may be represented as $CH_3—(CH_2)_n—COOH$. Numbering can be done either from the carboxyl end or from methyl end. Carboxyl carbon is the first carbon and methyl group carbon is the last one -omega. Still higher USFA can be synthesized from the above fatty acids with four, five and six double bonds. Examples for these are 20 carbon arachidonic acid (5, 8, 11, 14 eicosatetraenoic acid), 20 carbon 5, 8, 11, 14, 17 eicosapentaenoic acid (EPA) and 22 carbon 4, 7, 10, 13, 16, 19-docasohexaenoic acid (DHA) respectively. All these are physiologically very important compounds in humans and other mammals. EPA and DHA can be synthesized to a limited extent in humans and therefore have to be supplied in the diet. In general, the naturally occurring fatty acids are of even number of carbons but in bacteria and in ruminants because of their bacterial association can form odd chain and branched chain fatty acids.

A few very important physiological roles for triacylglycerols are: (a) contributing to obesity, thermal insulation and as a source of energy during hibernation. Stored lipid can be used as a source of energy during fasting or starvation. (b) In plants during the germination of oil rich seeds they act both as a source of energy and glucose as they can convert fatty acids into glucose. Since lipid or oil is of both nutritional and industrial importance increase in its quantity in the seeds is one of the important goals in agricultural biotechnology. Further since plants can synthesize EFA-which animals cannot- increasing their synthesis is another goal. (c) fatty acids play an important role in regulating the membrane fluidity. (In membranes it is not triacylglycerols but phospholipids which possess fatty acids). Under low temperatures USFA increase while under high temperature SFA increase. So nature of fatty acids play an important role in temperature adaptation. Agricultural biotechnologists are trying to improve cold tolerance by increasing the USFA content by manipulating the enzymes desaturases responsible for introducing the double bonds by modern molecular techniques. Further as plants alone can produce about 200 different fatty acids which have nutritional and industrial applications agricultural biotechnologists are manipulating both the quantity and quality of various oils/fatty acids by modern molecular techniques. (d) Increased fatty acid synthesis by the enzyme fatty acid synthase (FAS) is observed in some breast cancers. Hence using FAS inhibitors is a therapeutic approach. Hence the metabolism has a clinical significance. Similarly inhibitors of one of the two isoenzymes of diacylglycerol acyltransferase (DGAT-1) which is involved in triacylglycerol synthesis are also of clinical importance as they are exploited for the treatment of obesity.

On complete hydrolysis by lipases triacylglycerols liberate one molecule of glycerol and three molecules of fatty acid. The intermediates are diacylglycerol and monoacylglycerol. The glycerol molecule can be converted to glucose by all organisms but this constitutes only a tiny fraction, about 5 per cent of the Triacylglycerol. However, the fatty acid portion can be metabolised and utilised in different ways in different organisms. In animal systems including humans fatty acids are catabolised to acetyl-CoA by the successive removal of two carbons at a time by a process called beta oxidation. In case odd chain fatty acids are catabolised, the last one molecule formed is a three carbon propionyl CoA. Propionyl CoA can be metabolised to enter TCA cycle or to form glucose. Beta-oxidation pathway can take place both in mitochondria and peroxisomes/glyoxysomes. Acetyl CoA is metabolised in the TCA cycle to provide energy. During fasting or starvation acetyl CoA is used for the formation of ketone bodies - acetone, acetoacetyl-CoA and beta-hydroxybutyrate-which can be used as a metabolic fuel by the brain. Brain can use either glucose or ketone bodies as metabolic fuel. Plants and prokaryotes have the ability to convert fatty acids into glucose because of the presence of glyoxylate cycle which can channelize the acetyl CoA formed in beta-oxidation into glucose synthesis.(Discussed elsewhere). This process takes place in specialised organelles called glyoxysomes in plants, while in prokaryotes it is in the soluble portion. In eukaryotes the lengthening of fatty acids-elongation- and desaturation -creation of double bonds takes place in endoplasmic reticulum. Triacylglycerols are stored in lipid bodies or oleosomes in plants and in adipocytes in animals.

Hydrolysis of triacylglycerols

There are several lipases which can completely hydrolyse the triacylglycerol into free glycerol and three molecules of free fatty acids. There are specific lipases which act on:

(a) triacylglycerols liberating only one fatty acid and diacylglycerol, (b) diacylglycerols liberating one fatty acid and monoacylglycerol and (c) monoacylglycerols which liberate the last fatty acid and free glycerol. Accordingly they are termed as triacylglycerol lipase (TAG lipase), diacylglycerol lipase (DAG lipase) and monoacylglycerol lipase (MAG lipase) respectively. However, in humans and other mammals the pancreatic lipase can hydrolyse two fatty acids and liberate a monoglyceride. Whereas from oral cavity and gastric lipases can hydrolyse only one fatty acid liberating a diacylglycerol. pH optima of these lipases is different. Pancreatic lipase has an alkaline pH optima and others an acid pH optima. Acid and alkaline lipases are common in many organisms.

Fatty acid breakdown by Beta-oxidation

(a) General Introduction

It is called beta-oxidation as it is the beta carbon i.e. third from the carboxyl end—that is oxidized from CH_2 to C=O. As mentioned above on complete hydrolysis by lipases we get three molecules of fatty acids. The site of occurrence of the beta-oxidation pathway may vary with the length of the fatty acid. Short and medium chain fatty acids are oxidized in mitochondria while long chain fatty acids can be oxidized both in mitochondria and peroxisomes. Very long chain fatty acids 18 plus to 26 are oxidized exclusively in peroxisomes. In germinating seeds, fatty acids liberated from triacylglycerols undergo both beta-oxidation and glyoxylate cycle in glyoxysomes (considered as a specialised form of peroxisomes).

Reactions of beta-oxidations are not unfamiliar in view of our previous discussion of metabolic pathways. Recall how succinate is converted to oxaloacetate in TCA cycle. A dehydrogenation followed by hydration, and another dehydrogenation. A similar sequence of reactions occurs in the beta-oxidation of fatty acids. The above three reactions are preceded by the activation of fatty acids and ended with the cleavage of two carbon groups in the form of acetyl CoA. So the sequence of reactions in beta-oxidation of fatty acid (CH_3—$(CH_2)_n$—COOH) are: activation—dehydrogenation—hydration—dehydrogenation—cleavage. The above sequence of reactions are repeated (except activation—activating the fatty acid once is sufficient for its complete oxidation). In every round of the above reactions we get one molecule of two carbon compound -acetyl CoA. In the last round when four carbons are remaining we get two molecules of acetyl CoA. As mentioned above odd chain fatty acids give a molecule of 3 carbon propionyl CoA in the last round.

Fatty acids have to be transported into mitochondria. While short and medium chain fatty acids can be transported into the matrix directly as free fatty acids and then get activated, long chain fatty acids have to be first activated at the outer surface of the mitochondrial membrane and then enter mitochondria. (The classification of fatty acids into short, medium, long and very long seems to slightly vary with the author. However, I recommend short chain to be of 4 to 6 carbons; medium chain to be of >6 to 12, long chain >13 to 21 and very long chain 22 and above number of carbons. However, most of the naturally occurring important fatty acids are even numbered. Hence the above figures may also be quoted as simply 6-12, 14-20 and 22 and above).

(b) Pathway

(a) Activation of fatty acids means converting them into CoA derivatives:

Fatty acid + CoA + ATP → fatty acyl-CoA (also written as fatty acyl-SCoA) + AMP + PPi

The enzyme catalyzing the activation reaction is called differently as - fatty acyl CoA synthetase or fatty acyl CoA synthase or acetyl CoA ligase or thiokinase. This activation is required for the fatty acids to undergo subsequent reactions. However, the activation does not help in the transport of the long chain fatty acids into mitochondrial matrix. They have to be converted from CoA esters into carnitine esters (carnitine is 4-trimethyl amino 3-hydroxy butyrate). Carnitine ester formation is only to help in transport but for catabolism by beta- oxidation the fatty acid has to be in the form of CoA ester. So carnitine ester has to be reconverted to CoA ester of fatty acid. These reactions take place as follows:

(i) Carnitine + fattyacyl- CoA $\rightarrow$ fattyacyl- carnitine + CoA; This reation is catalyzed by carnitine acyltransferase-I (CAT-I) located on the outer surface of the inner membrane of mitohondria.

(ii) fattyacyl- carnitine + CoA $\rightarrow$ fatty acyl-CoA + carnitine. Fattyacyl- carnitine is transported across the inner membrane by a specific translocase-carnitine-acylcarnitine translocase. Further reconversion to CoA derivative is carried out by another enzyme -CAT-II. Carnitine formed above is transported back to conjugate again with another fatty acid molecule.

In peroxisomes or glyoxysomes no need of transport in the form of carnitine esters. However, the activation process unlike in mitochondria involves oxygen forming hydrogen peroxide (H_2O_2) a toxic metabolite which is broken down to water (H_2O) and oxygen (½ O_2) by the enzyme catalase found in the organelle. Glyoxysomes are found not only in higher plants but also in algae, fungi, yeast, and protozoa.

(b) the first dehydrogenation is carried out by the enzyme fatty acyl CoA dehydrogenase. There can be specific dehydrogenases for short, medium, long and very long chain fatty acids.

Fatty acyl-CoA + FAD $\rightarrow$ trans Δ-2 enoyl CoA + $FADH_2$

(c) the next step of hydration-addition of H_2O- is carried out by enoyl-CoA hydratase giving L-betahydroxyacyl-CoA. This product undergoes a dehydrogenation by the enzyme

L-betahydroxyacyl-CoA dehydrogenase:

L-betahydroxyacyl-CoA + NAD^+ $\rightarrow$ beta-ketoacyl-CoA + NADH + H^+;

(d) the last step involves the cleavage of a two carbon group in the form of acetyl CoA by thiolase enzyme; hence it is called thiolytic cleavage.

beta-ketoacyl-CoA + CoA → acetyl CoA + fatty acyl-CoA with n-2 i.e. two carbons less than the original number. Since the resulting shorter fatty acid is already in the form of fatty acyl-CoA, activation step is bypassed. The resulting fatty acyl CoA undergoes repeated rounds of the above mentioned reactions till it is completely broken down. i.e. if it is palmitic acid ($C_{15}H_{31}COOH$ or CH_3–$(CH_2)14$–COOH) it is completely broken down in 7 rounds of beta-oxidation giving 8 molecules of acetyl CoA. (in the last round two molecules of acetyl CoA are liberated by thiolase reaction. Hence number of acetyl CoA molecules are one more than the number of rounds of beta-oxidation). If it is stearic acid ($C_{17}H_{35}COOH$ or CH_3–$(CH_2)16$–COOH, we get 9 molecules of acetyl CoA after 8 rounds of beta-oxidation. Acetyl CoA formed as above enters the TCA cycle. Hence we get energy both because of acetyl CoA oxidation in TCA cycle and also because of oxidation of reducing power formed in the dehydrogenase reactions-$FADH_2$ and NADH + H^+.

In case of odd chain fatty acids, as mentioned above the last molecule is propionyl CoA which can enter TCA cycle or go for gluconeogenesis. This occurs as follows:

(e) Propionyl CoA + CO_2 + ATP → D-methylmalonyl CoA + ADP + Pi catalyzed by propionyl CoA carboxylase;

(f) D-methylmalonyl CoA is isomerised to L-methylmalonyl CoA by a racemase;

(g) L-methylmalonyl CoA is further isomerised to succinyl CoA by methylmalonyl mutase.

(h) Succinyl CoA formed as above can enter TCA cycle or can be metabolised towards glucose forming reactions.

Oxidation of USFA requires at least two additional steps. In the above mentioned process for SFA, after activation of fatty acids the first dehydrogenation creates a double bond between carbon 2 and carbon 3 also called alpha-beta unsaturation and these double bonds are in trans form. At other positions there is no unsaturation in SFA. Therefore, for USFA regular beta oxidation cycles take place till it encounters a fatty acid any double bond other than at the above position. So the combined action of two enzymes is required to create molecule to participate in regular beta-oxidation process. An enoyl CoA isomerase shifts the double bond from other positions to alpha- beta position in a trans form. Further additional double bonds are reduced by 2,4, -dienoyl CoA reductase. These two enzymes create an intermediate that can participate in the regular beta-oxidation process. Whenever an additional double bond is encountered in the molecule thereafter the combined action of the above two enzymes can generate an intermediate suitable for regular beta-oxidation process.

(c) Alpha and Omega oxidation of fatty acids

(a) Alpha-oxidation occurs at C-2 position rather than on carbon 3rd as in beta-oxidation. It removes a carboxy group from the terminus. It is of significance both in plants and animal systems. Though initially the fatty acid is metabolised by alpha-oxidation, subsequently it undergoes beta-oxidation. This reaction is aerobic and does not require any activation. Phytol which is a constituent of chlorophyll has Phytanic acid which undergoes alpha-oxidation followed by beta-oxidation. This process can take place both in plants and also in animals as phytol can come from the diet, not only plant based but also from milk and animal fats. Phytanic acid is initially oxygenated by a hydroxylase which in turn is decarboxylated by an oxidase giving pristanic acid. Deficiency of alpha-oxidation leads to a clinical disorder called Refsum's disease characterised by several neuronal and opthalamological abnormalities in children. These individuals should not consume any chlorophyll based or Phytol based diets.

(b) Omega Oxidation - Like alpha-oxidation it is also a minor pathway. The reaction initially involves an hydroxylase in the endoplasmic reticulum followed by an alcohol dehydrogenase and aldehyde dehydrogenase converting the omega carbon i.e. the last carbon-methyl group - to carboxylic acid and therefore we have a fatty acid with two carboxyl groups –one on either end of the fatty acid. This pathway is implicated in different human pathological states like diabetes, chronic alcoholic consumption etc.

(c) Any reduction or absence of peroxisomes can also effect the oxidation of long chain fatty acids and lead to clinical abnormalities.

9

Fatty Acid and Triacylglycerol Biosynthesis

Fatty acid Biosynthesis

(a) General Introduction

There are a huge number of fatty acids. Plants alone can produce about 200 different fatty acids. Hence what is discussed here is the synthesis of fatty acids which are of general occurrence in all organisms. The sub-cellular site of synthesis of fatty acids is chloroplast in case of plants and cytosol in other organisms. The enzymes involved exist as a multienzyme complex or a multifunctional polypeptide. In vertebrates, seven enzyme activities reside on the same polypeptide. Hence it is a large multifunctional polypeptide. In bacteria and plants, the seven enzyme activities exist as seven separate polypeptides. A variation in yeast is the seven activities are distributed in to two separate multifunctional polypeptides.

Recall the steps in beta-oxidation- activation-dehydrogenation-hydration-dehydrogenation-cleavage. During fatty acid synthesis in a way it is the opposite/reverse set of reactions that take place: Joining—reduction-dehydration-reduction and the cycle repeats.

However, there are important contrasts. In beta-oxidation in every round we get a molecule of two carbon acetyl CoA as a result of thiolytic cleavage. In fatty acid synthesis the intermediates are not in the form of CoA derivatives but as acyl carrier protein (ACP).

ACP also has an -SH group like in CoA. Hence, the joining is between one molecule of acetyl -ACP and one molecule of a three carbon compound, malonyl- ACP. However, during the joining reaction one carbon is removed from malonyl -ACP as CO_2. So the net result is, as if two molecules of two carbon groups (acetate units) are joined to increase the length by two carbons in every round of the fatty acid synthesis reactions. The liberated CO_2 can be used to carboxylate acetyl CoA to form malonyl CoA. The CoA derivatives are subsequently converted to ACP derivatives enzymatically. Acetyl CoA for fatty acid synthesis can come from the catabolism of carbohydrates and proteins. Though in organisms other than plants fatty acid synthesis takes place in cytosol, the acetyl CoA originates from the citrate -a TCA cycle intermediate- transported from mitochondria to cytosol.

(b) Pathway

(a) Citrate + CoA + ATP → oxaloacetate + acetyl CoA + ADP + Pi catalyzed by the enzyme ATP dependent citrate lyase.

(b) acetyl CoA + CO_2 + ATP → malonyl CoA + ADP + Pi, catalyzed by the enzyme acetyl CoA carboxylase.

The above formed CoA derivatives are converted to ACP derivatives as follows:

(c) acetyl CoA + ACP → acetyl-ACP + CoA, catalyzed by the enzyme acetyl CoA- ACP transacetylase.

(d) malonyl CoA + ACP → malonyl -ACP + CoA, catalyzed by the enzyme malonyl CoA- ACP transferase.

Joining or condensation of acetyl-ACP and malonyl-ACP then is catalyzed by 3-ketoacyl ACP synthase,

(e) acetyl-ACP + malonyl-ACP → 3-ketoacyl ACP + CO_2.

(f) reduction of 3-ketoacyl ACP using NADPH + H^+ (this is derived from OPP pathway. Hence the fatty acid synthesis pathway is linked to the production of reducing power by OPP pathway.) This reduction is catalyzed by beta-ketoacyl ACP reductase which reduces the carbonyl group at the beta position, giving 3-beta-hydroxyacyl-ACP and $NADP^+$.

3-ketoacyl ACP + NADPH + $H^+ \rightarrow$ 3-beta-hydroxyacyl-ACP + $NADP^+$.

(g) dehydration of 3-beta-hydroxyacyl-ACP to form trans delta 2 butenoyl-ACP and H_2O by 3-beta-hydroxyacyl-ACP dehydratase.

(h) a second reduction again by using NADPH + H^+ by enoyl-ACP reductase;

trans Δ 2 butenoyl-ACP + NADPH + $H^+ \rightarrow$ butyryl-ACP + $NADP^+$;

butyryl-ACP is a saturated ACP derivative (CH_3-CH_2-CH_2-CO-ACP), a four carbon derivative which participates in another round of the above mentioned reactions "e" to "h".

If the above cycle is repeated seven times we get a 16 carbon saturated fatty acid as ACP derivative i.e. palmitoyl-ACP.

(i) palmitoyl-ACP is hydrolysed by a thioesterase to form free palmitic acid and ACP. CoA, ACP and NADP formed in the above reactions are recycled.

Generally fatty acid synthesis stops at palmitate (16 carbons). However, to some extent it can form the next higher stearate (18 carbons). In case stearate is not formed from the above series of reactions, palmitate has to be elongated in the endoplasmic reticulum by the enzyme bearing the trivial name elongase. Further elongation to still longer lengths is also carried out as above. Elongation and formation of USFA takes place in the form of CoA derivatives. Both the processes take place in endoplasmic reticulum. For elongation, as above malonyl CoA is utilised and unsaturation i.e. creation of additional double bonds is catalyzed by specific desaturases using NAD^+ and forming NADH + H^+.

As mentioned above humans can synthesize only mono unsaturated fatty acids i.e. can introduce only one double bond. The enzyme oleate desaturase catalyzes the dehydrogenation of stearate to oleate which is also 18 carbon fatty acid but with a double bond between carbons 9 and 10. (For EFA-linoleic and linolenic acids and other PUFA see discussion above).

Triacylglycerol Biosynthesis

Triacylglycerol Biosynthesis - two pathways are there -

(a) Glyerol-3-phosphate pathway

(a) Glycerol-3-phosphate + fatty acyl CoA → lysophosphotadic acid + CoA

Enzyme is Glycerol-3-phosphate acyl transferase. The fatty acid is esterified at the first position (R) of the glycerol.

Glycerol 3-phosphate is synthesized by reducing dihydroxyacetone phosphate (DHAP), formed in glycolysis by the action of glycerol-3-phosphate dehydrogenase. Hence gluconeogenic amino acids can lead to the formation of glycerol 3-phosphate. This is just to indicate how the metabolism is interlinked. It can also be formed by the phosphorlyation of free glycerol formed in the complete hydrolysis of triacylglycerols by glycerol kinase-

Glycerol + ATP → glycerol-3-phosphate + ADP

(b) lysophosphatidic acid + fatty acyl CoA → phosphatidic acid (this fatty acid at second position (R^1) acylglycerol phosphate acyl transferase (AGPAT))

(c) phosphatidic acid +H_2O → diacylglycerol +Pi (phosphatidic acid phosphohydrolase (PAP))

(d) diacylglycerol + fatty acyl CoA → triacylglycerol + CoA; (Diacylglycerol acyltransferase (DGAT)).

(b) Monoacylglycerol pathway

(a) Monoacylglycerol + fattyacyl CoA → diacylglycerol + CoA (monoacylglycerol acyl transferase).

(b) diacylglycerol + fatty acyl CoA → triacylglycerol + CoA (Diacylglycerol acyltransferase (DGAT).

The reaction catalyzed by DGAT is considered the terminal and only committed step in triglyceride synthesis and to be essential in humans in the adipocytes. There are two isoenzymes of DGAT, DGAT1 and DGAT2.

Salient regulatory details of Fatty acid and Triacylglycerol Biosynthesis and Break down

As mentioned above regulation can occur at many steps. Further fatty acid oxidation and fatty acid synthesis are reciprocally regulated. Both are not active at the same time otherwise it becomes a futile cycle. Hence both processes are discussed simultaneously. Salient regulatory features are:

(a) In mammals lipases can be activated by the hormones, glucagon, epinephrine, nor-epinephrine and adrenocorticotropic hormone (ACTH). This activation takes place because of a covalent modification by phosphorylation by cAMP mediated protein kinase A (PKA). (Read discussion under signal transduction). Insulin causes dephosphorylation and inactivates the enzyme. It is the pancreatic lipase and TAG lipase that is subjected to hormonal regulation. Details about other lipases not known. There seems to be isoenzymic control of DAG transferases involved in triacylglycerol synthesis.

Plant lipases also are regulated by plant hormones. GA_3 increases while ABA inhibits Plant lipases require calcium and zinc for their activation.

The reaction catalyzed by DGAT is considered the terminal and only committed step in triglyceride synthesis and to be essential in humans in the adipocytes. There are two isoenzymes of DGAT, DGAT1 and DGAT2.

(b) When energy supply is high fatty acid/lipid synthesis is favoured. So higher energy levels as indicated by ATP, NADH + H^+ or acetyl CoA inhibit beta-oxidation enzymes and promote fatty acid synthesis. NADH + H^+ inhibits 3-hydroxy acyl CoA dehydrogenase and acetyl CoA inhibits thiolase.

(c) Acetyl CoA carboxylase is feedback inhibited by palmitoyl CoA or other longer chain fatty acyl CoA's. Enzyme is inactivated by covalent modification by phosphorylation. It is serine residue in the enzyme that is phosphorylated. This phosphorylation is mediated by the hormones glucagon and epinephrine in mammalian systems. Phosphorylation inactivates the enzyme by causing the dissociation of the active octamer form of the enzyme into monomers. The protein kinase that phosphorylates—is not cAMP dependent but AMP dependent. However, cAMP has a role as it activates "AMP dependent protein kinase" by phosphorylation by another kinase which is cAMP dependent. AMP-protein kinase is inactivated by dephosphorylation. Insulin can reverse the process. Insulin activates protein phosphatase 2 A and causes dephosphorylation of the above enzymes and promotes polymerisation of the monomers of acetyl CoA carboxylase into active octameric form of the enzyme.

High levels of AMP i.e. a low energy charge inhibits the enzyme. High AMP activates TCA cycle and decreases fatty acid synthesis.

(d) Role of citrate- Citrate can allosterically activate the enzyme acetyl CoA carboxylase. It inhibits PFK-1 thereby reducing the flow of carbon in the glycolytic pathway. Citrate also can promote the polymerisation of monomers of acetyl CoA carboxylase making it active. As mentioned above high AMP inhibits TCA cycle. It can inhibit ICDH leading to the accumulation of citrate in mitochondria. Citrate then can move into cytosol and increases fatty acid synthesis as it can act as a substrate for ATP dependent citrate lyase providing acetyl CoA for fatty acid synthesis. ATP dependent citrate lyase is activated by insulin.

(e) Role of palmitoyl CoA- Palmitoyl CoA inhibits fatty acid synthesis in many ways. One way is mentioned above. Other ways are: (i) it inhibits citrate translocation into cytosol by inhibiting translocase. Therefore, no fatty acid synthesis, (ii) inhibits the OPP pathway enzyme G6PDH thereby reducing the supply of the reducing power NADPH + H^+ needed for fatty acid synthesis, (iii) It can also cause the dissociation of the enzyme subunits of acetyl CoA carboxylase making it inactive, (iv) it activates cAMP dependent protein kinase. How this leads to inactivation of acetyl CoA carboxylase is mentioned above.

(f) Malonyl CoA inhibits beta-oxidation by inhibiting CAT-1. So fatty acids cannot enter mitochondria to get catabolised (applicable to long chain fatty acids). In plants there can be an isoenzymic control as there are three forms of ketoacyl ACP synthase.

(g) Excess of PUFA inhibit the transcription of genes involved in fatty acid synthesis.

(h) From the above mentioned points it can be stated that low energy, low insulin and starvation promotes beta-oxidation.

Pathway manipulation and applications

Though the above regulatory mechanisms are known, manipulation of lipid metabolism is carried out by modern molecular techniques (discussed under genetic regulation) rather than by an understanding of conventional regulation of metabolic pathways as it was done in case of manipulating amino acid biosynthetic pathways. Of course knowledge of biosynthetic pathways to select relevant genes has been utilised. Mostly these manipulations have been attempted in plants because of the nutritional significance and as they can be a source of large scale supply of starting material for industrially relevant products like soaps, detergents, shampoos (lauric acid a 12 carbon fatty acid is obtained from coconut and palm kernel) lubricants, adhesives, plastics, paints, varnishes, Biodiesel or Biofuel (methyl esters of plant oils) etc. Some examples which are of biotechnological relevance are:

(a) Erucic acid (22:1 Δ13 or omega-9) is a major constituent of rape seed oil. It is harmful as it can cause heart disease and hence its content needs to be reduced. A low erucic acid oil variety called canola has been developed. As it is synthesized from oleic acid its conversion to erucic acid is blocked. Erucic acid can be as high as 54 and 42 per cent in rape seed and mustard oil respectively. By blocking the further metabolism of oleic acid into erucic acid, the content of erucic acid could be reduced to as low as 2 per cent -also called LEAR oil. However, it has industrial applications as surfactants, lubricants and biodiesel for which high levels are required from the plants. Over expression of beta-ketoacyl synthase in rape seed and elongases has led to a high erucic acid levels (HEAR).

(b) Anti-sense suppression of oleoyl-desaturase in soyabean causes an increase in oleic acid accompanied by an increase in saturated fatty acids. This is useful for cooking purposes. More stable at high temperatures during frying.

(c) Anti-sense suppression of Steoryl-ACP desaturase leads to 30 to 40 per cent increase in saturated fatty acids which are useful in margarine preparation.

(d) Increasing acetyl CoA carboxylase increases oil content in Brassica by 5 per cent.

(e) Lauric acid content can be increased by 40-50 per cent by cloning a gene for medium chain (12:0) acyl-ACP thioesterase and over expressing it. Acyl-ACP thioesterases for specific chain lengths are available and these can be exploited to produce fatty acids of different chain length.

(f) Ricinoleic acid (12OH-18:1 Δ9) is present in seeds of castor beans and is a useful material for the synthesis of nylon, lubricants, plastics, cosmetics, hydraulic fluids, etc. However, the castor bean crop has some disadvantages like low yield, presence of the toxin ricin etc. Hence, it is advantageous to reproduce this fatty acid from other plants. Experimental demonstration has been done by cloning the oleate hydroxylase (mono-oxygenase) gene into tobacco.

(g) Polyhydroxy butyrate finds utility as a biodegradable plastic. It is naturally produced by the microorganism *Alcaligenes eutrophus* from acetyl CoA. But producing this by microbial fermentation in large quantities is expensive. Hence producing them in plants is attempted. Three genes responsible-phbA (3-ketothiolase), phbB (acetoacetyl CoA reductase) and phbC (polyhydroxy butyrate synthase) are cloned into the model plant Arabidopsis. The genes were successfully expressed in chloroplasts and PHB accumulated to the extent of 14 per cent by dry weight of leaves without affecting chloroplast functions.

(h) Cloning and over expression of microsomal desaturases, delta-12 desaturase and delta-6 desaturase has resulted in increased formation of linoleic acid and gamma-linolenic acid respectively.

10

Cholesterol Biosynthesis and Regulation

(a) General Introduction

Cholesterol is classified as a lipid molecule and has the basic steroid ring-cyclopentanoperhydrophenanthrene- possessing the molecular formula $C_{27}H_{45}OH$. It is an essential component of all animal cell membranes -nearly 30 per cent of the membranes- and controls the membrane integrity and fluidity. Nearly 30 per cent of the total human body cholesterol is found in the brain and all of it is locally synthesized i.e. in the brain itself. Cholesterol is also a precursor for all steroid hormones *viz.* estrogen, progesterone, testosterone, cortisol, aldosterone etc. and Vitamin D. It is also a precursor of bile acids involved in the emulsification of fats in the digestive tract and absorption of fat molecules as well as the fat-soluble vitamins, A, D, E and K. Cholesterol is absent from prokaryotes barring exceptions. Plants also synthesize small amounts of cholesterol and also related compounds called phytosterols e.g. campesterol, sitosterol and stigmasterol. These phytosterols can be obtained from several seeds and has beneficial effects on human health in lowering the cholesterol levels. Plants also have cholesterol in their membranes.

(b) Pathway

Humans can endogenously synthesize cholesterol and also obtain from diet. High levels of cholesterol can lead to heart disease and stroke. Hence its level needs to be controlled. It is synthesized in the liver. The biosynthetic pathway involves a very large number of steps starting from mevalonate—hence it is a highly energy requiring pathway- and mevalonate itself is synthesized from acetyl CoA in three steps. (a) Two molecules of acetyl CoA combine to form acetoacetyl CoA by the thiolase enzyme. Thiolase enzyme in beta-oxidation of fatty acids normally breaks down acetoacetyl CoA group into two molecules of acetyl CoA. Hence the above reaction is called as a reversal of thiolase reaction:

(a) acetyl CoA + acetyl CoA → acetoacetyl CoA + CoA.

(b) acetoacetyl CoA so formed reacts with another molecule of acetyl CoA to form 3-hydroxy-3-methyl-glutaryl CoA by the action of hydroxymethylglutaryl (HMG) CoA synthase;

acetoacetyl CoA + acetyl CoA → 3-hydroxy-3-methyl-glutaryl CoA + CoA

(c) The third and the major regulatory step involves the formation of mevalonate by the action of HMG- CoA reductase, which is a component of the membrane of endoplasmic reticulum.

3-hydroxy-3-methyl-glutaryl CoA + 2 NADPH + H^+ → mevalonate + 2 $NADP^+$ + CoA.

(c) Regulation

Regulation of cholesterol biosynthesis is very important in view of the points mentioned above. Cholesterol in turn regulates the levels of so many compounds. Though the pathway has a large number of steps only one enzyme/reaction -HMG CoA reductase-is considered to be the major controlling point and is regulated in many ways. This enzyme is a target for anti-cholesterol drugs, broadly referred to as -statins e.g. lovastatin, simvastatin, atorvastatin, mevastatin etc. So there appears to be lying undiscovered the regulation of many steps/enzymes. However, presently there are many ways of regulation identified which are as follows:

(a) HMG-CoA reductase activity itself is regulated by covalent modification by phosphorylation. The enzyme is inactive when it is phosphorylated-serine 871 is phosphorylated -by the enzyme AMP dependent protein kinase. The donor of phosphate group for protein kinase is ATP. Its activity is restored by dephosphorylation by the enzyme HMG-CoA reductase phosphatase. The kinase itself can exist in active and inactive forms. The kinase is in an active form when phosphorylated by another kinase—HMG-CoA reductase kinase kinase- at the expense of ATP and becomes inactive if phosphate group is removed. The removal of the phosphate group -dephosphorylation -is catalyzed by HMG-CoA reductase kinase phosphatase.

(b) Cholesterol -the end product- itself is a feedback inhibitor of the already existing enzyme HMGR. It also causes rapid degradation of HMG-CoA reductase by ubiquitin mediated protein degradation (See notes under ubiquitin degradation for the mechanism). Increased dietary intake of cholesterol decreases the endogenous synthesis of cholesterol. Here the reduction does not occur because of the dietary cholesterol directly inhibiting the HMGR but otherwise regulating its biosynthesis.

(c) Cholesterol levels regulate its biosynthesis by the action of a transcription factor –SREBP (sterol regulatory element binding protein)- which binds to the 5' end of the HMGR gene. Cholesterol levels regulate the transcription of HMGR gene. The 5' portion of the gene is called sterol regulatory element (SRE). SREBP is bound normally to the membranes of endoplasmic reticulum (ER). It can also be found in the nuclear membrane. SREBP in the ER is associated with another protein called SREBP cleavage protein abbreviated as SCAP is also found in ER. When cholesterol levels are high, SCAP binds to cholesterol, undergoes a conformational change and binds to another protein Insig (insulin induced gene). In this altered conformation and status, it binds to SREBP forming a SREBP-SCAP complex. In this state SREBP remains in the ER and cannot contribute to the transcription of the HMGR gene. But when cholesterol level decreases the conformation of SCAP changes causing the release of bound SCAP-SREBP complex from ER. The complex so released moves to golgi vesicles where it is proteolytically cleaved twice and releases the N-terminal domain into the cytosol. It is only during the second cleavage that the N-terminal domain is released. The N-terminal domain then moves from cytosol into nucleus, binds to SRE of not only HMGR gene but also other genes in the cholesterol biosynthetic pathway and increases their transcription. N-terminal domain has a short half-life and is degraded at a faster rate by ubiquitin mediated degradation. When cholesterol levels are increased SCAP-SREBP remains bound to the membrane of ER and the cleavage of SREBP does not occur and hence no transcription is facilitated.

(d) Lanosterol is an intermediate in the biosynthesis of cholesterol starting from mevalonate and is also considered as a precursor of cholesterol. Lanosterol can accumulate in the membrane of ER. When it is accumulated the N-terminal region of HMGR protein binds to another protein Insig (insulin induced gene). This binding signals ubiquitin mediated degradation of HMGR and the enzyme is rapidly degraded. Thereby the half-life of the enzyme decreases from > 12 hrs to less than one hour.

(e) HMG CoA reductase is regulated by several hormones; Insulin, glucagon, epinephrine, glucocorticoids and Thyroid hormones. Insulin leads to a decrease in cAMP, which in turn activates cholesterol synthesis while glucagon and epinephrine, by increasing the level of cAMP, inhibit cholesterol synthesis. Thus, HMGR activity is also controlled by the cAMP signaling pathway. cAMP-dependent protein kinase, PKA, phosphorylates phosphoprotein phosphatase inhibitor-1 (PPI-1) causing its activation. PPI-1 can then inhibit the activity of phosphatase 2C (PP2C) and PP2A (also called HMGR phosphatase) which remove phosphates from AMPK and HMGR, respectively. Thus, AMPK remains phosphorylated and active, while HMGR in the phosphorylated state is inactive. Hence presence or absence or drastic alterations in the levels of the above mentioned hormones can regulate cholesterol biosynthesis. The activity of HMGR is increased by Thyroid hormones and decreased by glucocorticoids.

(f) Translation of HMGR mRNA can be inhibited by several intermediate non-steroid metabolites in the biosynthetic pathway of cholesterol from mevalonate.

(g) Cholesterol can be present in a cell in free form or in a esterified form as acyl CoA esters the reaction being catalyzed by acyl CoA cholesterol acyl transferase and the ester can be hydrolysed by cholesteryl ester hydrolase. It is also bound to lipoproteins. There can be influx or efflux of cholesterol into and out of the cell. Hence the levels of cholesterol in a cell are determined by

(i) extent of esterification and its hydrolysis, (ii) extent of influx and efflux, (iii) extent of its utilisation for the synthesis of other steroid hormones, bile acids, Vitamin D and related compounds.

(h) As mentioned above plants also synthesize small amounts of cholesterol and is found in membranes and also synthesize related compounds called phytosterols. These phytosterols can be obtained in the diet from several seeds and has beneficial effects on human health in lowering the cholesterol levels in humans. Hence they can be used as supplements to food items to lower the cholesterol. Many external stimuli like light, plant growth regulators, biotic stresses, herbicides, wound inducement can cause an increase in the transcription of HMGR.

Part - IV

Regulation of Amino Acid Metabolism

11

General Introduction

All the amino acids are synthesized from the carbon precursors arising from glycolysis, TCA cycle and OPP pathway. All the amino acids in addition to being the components of proteins they serve as precursors for the synthesis of numerous other metabolites. So amino acid biosynthetic pathways are interconnected with several other pathways. Further biosynthesis of several amino acids is energetically very expensive. Hence perfect regulation is required for the biosynthesis of amino acids. Humans suffer from many inborn errors of amino acid metabolism. Plants have a huge number of non-protein amino acids whose role and regulation remains an important area of research to be explored.

Humans and other mammals cannot synthesize all the 20 amino acids. All the 20 amino acids can be synthesized only by plants and microorganisms. Amino acids which cannot be synthesized by mammals but have to be supplied in diet. Those amino acids which can be synthesized by them are called non-essential amino acids (NEA) and those that need to be supplied exogenously in diet are called essential amino acids (EAA). In humans essential amino acids are: Methionine, Threonine, Tryptophan, Valine, Isoleucine, Leucine, Phenylalanine, Histidine and Lysine. Arginine is required for juveniles as they cannot synthesize them in required amounts. Tyrosine can be derived from Phenylalanine and so is not considered as an essential amino acid. Similarly, Cysteine can be derived from Methionine. (Students should know that in some text books cysteine biosynthesis may be shown from serine on a particular page. However, the first reaction from serine in the biosynthesis of cysteine requires L-homocysteine which is derived from methionine. Hence both methionine and serine are precursors for the biosynthesis of cysteine, methionine providing the sulfur and serine providing the carbon skeleton). That is why some classify Tyrosine and Cysteine as semi-essential as their supply in diet can spare some amount of Phenylalanine and Methionine required in the diet. The other amino acids are termed as non-essential as they can be synthesized by mammals. They are: Alanine, dicarboxylic amino acids i.e. aspartate and glutamate, amides i.e. Asparagine and Glutamine, Proline and Serine. Cysteine and Tyrosine can be added as they can be synthesized within the body from methionine and Phenylalanine respectively. (It should be kept in mind that the essentiality criteria are based on the ability or inability of the mammals to synthesize the different amino acids. Otherwise all 20 amino acids are absolutely required and indispensable for any organism as they are required for protein synthesis and several other metabolic pathways). Apart from the nutritional criteria amino acids can also be classified on the basis of their carbon metabolic precursor and the initial amino acid used for the synthesis of other amino acids into different biosynthetic families. Accordingly we have,

(a) Glutamate family i.e. amino acids synthesized from glutamate *viz.* glutamine, proline and arginine. Glutamate is derived from the carbon precursor alpha-ketoglutarate.

(b) Aspartate family i.e. amino acids synthesized from aspartate *viz.* asparagine, lysine, threonine and methionine. Aspartate is derived from the carbon skeleton oxaloacetate.

(c) Aromatic family *viz.* Phenylalanine, Tyrosine and Tryptophan. In this case the common carbon precursors for all the three aromatic amino acids are phosphoenol pyruvate and erythrose 4-phosphate.

(d) Branched chain amino acids family *viz.* isoleucine, leucine and valine. The carbon precursors for these amino acids are pyruvate and threonine derived from aspartate family. Hence they can also be termed as pyruvate family. Further alanine also is synthesised from pyruvate. Isoleucine can be justified to be discussed under aspartate family as a continuation of threonine metabolism. Similarly, cysteine also as a continuation of methionine metabolism.

(e) Serine family i.e. amino acids derived from serine *viz.* glycine and cysteine. Serine in turn is derived from 3-phosphoglycerate.

(f) Histidine stands alone as it is the only amino acid synthesized from the carbon precursor Ribose -5 Phosphate. Some books may write as PRPP (phosphoribosyl pyrophosphate which is formed from R5P). All the carbon precursors as mentioned above can be derived from either glycolysis or TCA cycle and OPP pathway. A biosynthetic pathway may constitute either essential or non-essential or both types of amino acids.

As only plants and microorganisms have the ability to synthesize the essential amino acids required by mammals, the details of their biosynthesis and regulation have been studied in these systems. And it is these details you find in general biochemistry books. Regulation of essential amino acids assumes importance as humans and other mammals are dependent on them and thus need to be produced in large amounts -primarily in the form of proteins and as free amino acids- by edible plant portions and by these amino acids if made to be produced in large amounts by microorganisms can be used as amino acid supplements. Proteins rich in essential amino acids are said to be of high biological value. After ethanol, antibiotics, amino acids constitute the third major commercial product important in microbial biotechnology. In 2015, worldwide market for amino acids was estimated to be 11.6 billion dollars. Both essential and non-essential amino acids have commercial applications in different sectors, like therapeutics, human and animal nutrition, cosmetics and flavour enhancers etc. One classical example of a non-essential amino acid is glutamate which is sold in tons as flavour enhancers in Chinese foods. Added as sodium glutamate the flavour is referred to as Umami. In microorganisms it is common to have biosynthetic operons for amino acid biosynthetic pathways and regulation at transcriptional level of polycistronic mRNA's. They are also regulated by attenuation and stringent response. In plants amino acid biosynthesis takes place considerably in chloroplast and many enzymes required for biosynthesis are coded by nuclear genes and hence are transported from cytosol. In animal systems non-essential amino acids can be synthesized in cytosol.

12

Biosynthesis and Regulation of Glutamate Family Amino Acids (Glutamate, Glutamine, Proline and Arginine)

Glutamate

It can be synthesized from alpha-ketoglutarate by the transamination reaction. Transaminases or amino transferases catalyze reversible transamination reactions. By these type of reactions glutamate can be formed mainly in three reactions. Further glutamate can also be formed under special conditions from alpha-ketoglutarate by glutamate dehydrogenase reaction and particularly by glutamate synthase (GOGAT-glutamine oxoglutarate amidotransferase) in the process of ammonia assimilation during biological nitrogen fixation in legumes and in the reassimilation of released ammonia during photorespiration and nitrate assimilation. The *in vivo* contributions of these four enzymes towards the formation of glutamate is not clear. However, it can be speculated that it all depends on the metabolic demands and the extent of the competing pathways operating in different directions.

(a) alpha-ketoglutarate + aspartate ↔ Glutamate + oxaloacetate. Since it is a reversible reaction it can also be considered as a biosynthetic reaction for the formation of aspartate. The direction in which it operates depends on the metabolic requirements.

The enzyme is called Glutamate-oxaloacetate transaminase (GOT or also as SGOT as it is routinely assayed in serum in clinical diagnosis) or Glutamate-aspartate transaminase or simply aspartate transaminase. In animal systems it has a considerable basal level expression and is further increased by gluocorticoids and cAMP but inhibited by insulin. It has cytosolic and mitohondrial isoforms which are differentially expressed. In plants the existence of isoenzymes in different subcellular compartments plays a critical role in regulating carbon and nitrogen pools among different compartments. GOT isoenzymes have been identified in cytosol, mitochondria, peroxisome, and chloroplasts of plants. Apparently there is a considerable exchange of carbon and nitrogen metabolites among these compartments. The enzyme has apparently an important role in transporting aspartate in C4 plants.

(b) alpha-ketoglutarate + alanine $\leftrightarrow$ Glutamate + pyruvate. As stated above this reaction may also be considered as a biosynthetic reaction for the biosynthesis of alanine. The enzyme is called Glutamate-pyruvate transaminase (GPT or also as SGPT as it is routinely assayed in serum in clinical diagnosis along with the above enzyme SGOT) or Glutamate-alanine transaminase or simply alanine transaminase.

In animal systems it is inhibited by cysteine and vitamin B6 derivatives. Androgens increase GPT activity and androgen response elements (ARE) have been identified in the promoter regions of the genes coding for GPT. There also exists an isoenzymic control as two isoforms with differential expression in different tissues has been observed. In wheat plants four isoforms are found and are differentially expressed and the activity varies in response to abiotic stresses, oxygen availability and nitrogen availability. While light upregulates, nitrogen deficiency downregulates GPT at transcriptional level.

(c) Glutamate + Glyoxylate $\leftrightarrow$ alpha-ketoglutarate + Glycine. This reaction catalyzed by glutamate- glyoxylate transaminase is an important reaction in photorespiration. However, the reaction in photorespiration proceeds towards the formation of glycine and then subsequently serine.

(d) NH_4^+ + alpha-ketoglutarate + $NADPH^+$ $\leftrightarrow$ Glutamate + $NADP^+$ + H_2O.

This enzyme (Glutamate dehydrogenase-GDH) is important for both synthesizing and catabolizing glutamate. In general, it is implicated in catabolism. But the possibility in the anabolic direction also exists. The enzyme can also utilize NAD^+. It appears that the specificity varies depending on the organism. Mammals can use either NAD^+ or $NADP^+$ while plant and microorganismal GDH is specific for either NAD^+ or $NADP^+$. Glutamate can participate in transamination reaction and also is a substrate for Glutamate family of amino acids. Further it can serve as an anaplerotic reaction feeding alpha-ketoglutarate to the TCA cycle. In case of humans when calorie intake is low alpha-ketoglutarate supplied as above can help generate ATP.

GDH is allosterically inhibited by GTP, GDP, Palmitoyl CoA and divalent zinc ions. It is allosterically activated by ADP, leucine, isoleucine and valine. Rationale for the above metabolites acting as activators or inhibitors is not clear. Enzyme is also regulated by covalent modification. Its activity is increased by ADP ribosylation. In mice GDH is inhibited by its own substrate glutamate when glutamate levels increase beyond a limit.

(e) In plants and microorganisms glutamate can also be formed from glutamate synthase (don't confuse with glutamine synthetase-GS) by the reaction:

alpha-ketoglutarate + glutamine + NADPH + H^+

$\leftrightarrow$ two molecules of Glutamate + $NADP^+$.

This enzyme is also called glutamate synthase/glutamine oxoglutarate amido transferase (GOGAT). As mentioned above this reaction is very important in the process of photorespiration and biological nitrogen fixation and nitrate assimilation. (Don't confuse between Glutamate synthase and Glutamine synthetase)

Glutamine

It plays an important role in nitrogen metabolism. It can donate its amino group from the side chain amide and become glutamic acid and glutamic acid in turn can donate another amino group to and get converted to a keto acid-alpha-ketoglutaric acid. Both amino acids participate in several reactions in different metabolic pathways.

Glutamine is formed from Glutamate by the action of glutamine synthetase (GS). The free ammonia liberated on oxidative deamination reactions or direct deamination reactions or free ammonia formed from nitrate or by direct assimilation of ammonia or formed in biological nitrogen fixation can be linked to Glutamate at the expense of ATP giving glutamine. Reaction may be represented as:

Glutamate + NH_4^+ + ATP $\rightarrow$ Glutamine + ADP + Pi.

The reaction is so important that it is regulated in all organisms. It is studied in greater details in *E.coli* which are discussed. Both glutamine and glutamate can also release free ammonia by the action of glutaminase and glutamate dehydrogenase (GDH). As GDH is a reversible reaction it may also synthesize glutamate under special conditions, though usually it works in a catabolic direction.

The enzyme GS is oligomeric consisting of 12 identical subunits and is regulated both allosterically and by covalent modification. The purpose of its regulation is to ensure supply of glutamine when it is low and decrease its supply when it is in surplus.

(a) GS is subjected to cumulative feedback inhibition. It is inhibited by several metabolites, each inhibiting partially. One metabolite inhibiting the enzyme does not interfere with the inhibition by another metabolite. So the enzyme has as many sites for the binding of metabolites which act as allosteric inhibitors. The presence of all metabolites will completely inhibit the enzyme. In all there are nine inhibitors of which six are considered as end products of glutamine metabolism and hence termed as end product inhibitors *viz.* histidine, tryptophan, Carbamoyl phosphate (synthesized by Carbamoyl Phosphate Synthetase II involved in pyrimidine biosynthesis), glucoseamine -6-phosphate, AMP and CTP. Three amino acids alanine, serine and glycine -their high levels- reflect the status of amino acid metabolism.

(b) Covalent modification- Covalent modification can further increase the susceptibility of the enzyme to allosteric inhibitors. Hence the enzyme can be inhibited by adenylation of a 397 tyrosine residue. This involves addition of AMP to tyrosine by adenylyl transferase. The donor of AMP is ATP and PPi is liberated. Each sub-unit can be adenylated.

Hence 12 adenylations can take place and the enzyme become progressively inactive with the successive addition of every adenyl group. So adenylation inactivates the enzyme. Removal of adenyl group-deadenylation-makes the enzyme more active. Both adenylation and deadenylation are carried out by the same polypeptide. These covalent modifications are in turn dependent on another regulatory protein called PII. This protein also can be covalently modified -not by adenylation—but by uridylation i.e. addition of UMP residues to specific tyrosine residue on the protein. The reaction is catalyzed by uridylyl transferase. The donor for UMP is UTP and PPi is liberated during the reaction. Both uridylation and deuridylation of PII is carried out by the same polypeptide. That is both adenyl transferase and also uridyl transferase are bifunctional polypeptides. Whether adenylation or deadenylation takes place depends on the PII status. When the adenylyl transferase forms a complex with uridylated PII(PII-UMP) it causes removal of adenyl groups i.e. it causes deadenylation. So glutamine synthetase is more active. If adenylyl transferase forms a complex with deuridylated PII, then this complex stimulates adenylation. So the enzyme glutamine synthetase becomes less active. PII protein in addition to regulating the activity of GS can also regulate the amounts of GS by increasing or decreasing the transcription of glutamine synthetase gene. PII in uridylated state increases the transcription of GS gene and in deuridylated state decrease the transcription of GS gene. Additional regulation occurs by allosteric regulation of Uridyl transferase. This enzyme is stimulated in the direction of uridylation by alpha-ketoglutarate and ATP and uridylation is inhibited when the enzyme is bound to glutamine and Pi. In the above process PII may combine with some more proteins.

Proline

Proline plays an important role in conferring resistance to abiotic stresses (particularly water and salinity) in plants and also in microorganisms. It is accumulated under such stress conditions as an osmolyte. Though it is accumulated in high concentration in the cell it is not harmful and hence it is referred to as compatible solute. (Students should keep in mind that proline is not the only osmolyte/compatible solute accumulated under stress conditions. There are some more).

Proline in addition to acting as an osmolyte also stabilizes membranes and biomolecules and acts as an anti-oxidant. Further it acts as a storage amino acid as it can be reutilised when once stress is relieved. Hence regulation of its biosynthesis and degradation is very important for the organism. Proline is synthesized form glutamate which is the major pathway. In plants it can also by synthesized from ornithine. Regulation from glutamate is discussed here.

The important step is the conversion of glutamate to delta-1-pyrroline 5-carboxylate (P5C) in three sub-steps- two of which are enzymatically catalyzed and the third one being a non-enzymatic spontaneous cyclisation- by the bifunctional enzyme delta-1-pyrroline 5-carboxylate synthetase (P5CS):

(a) Glutamate + ATP → glutamyl-gamma-phosphate + ADP; Enzyme here is gamma-glutamyl kinase (GGK),

(b) glutamyl-gamma-phosphate + NADPH+ H^+ → glutamic - gamma-semialdehyde (GSA) + $NADP^+$ + Pi; This reaction is catalyzed by GSA dehydrogenase. Both GGK and GSA dehydrogenase activities reside on the same protein together referred to as P5CS. The subsequent reaction is spontaneous cyclisation of GSA to P5C.

GGK is feedback inhibited by proline. Proline dehydrogenase is the degradative enzyme which converts proline back to P5C. There is a strict reciprocal regulation between the activities and amounts of P5CS and Proline dehydrogenase (as indicated by the abundance of transcripts for both the enzymes) during stress and after stress recovery. When once stress is relieved P5CS decreases and proline dehydrogenase increases. In plants the hormone abscisic acid (ABA) seems to be involved in increasing P5CS activity.

Arginine

(a) Importance

Arginine, nutritionally is considered as inadequate only for juveniles and not for adults. Hence it is not classified as an essential amino acid. Since arginine formed in the urea cycle in mammals is hydrolyzed to ornithine and urea by arginase, it remains to be established how exactly the total requirement of arginine for the organism is satisfied. Arginine in different organisms is a precursor for important metabolites like polyamines, Nitric oxide, etc. In plants it is an important form of nitrogen storage.

(b) Pathway

In different organism arginine can be synthesized from either glutamate or ornithine. Their reactions are as follows:

(a) Glutamate + acetyl CoA → CoA + N-acetylglutamate, catalysed by N-acetylglutamate synthase (NAGS);

In the next step N-acetylglutamate is phosphorylated at the expense of ATP and subsequently reduced to form N-acetylglutamate semialdehyde;

(b) N-acetylglutamate + ATP N- acetylglutamylphosphate + ADP, catalysed by N-acetylglutamate kinase (NAGK);

(c) N-acetylglutamylphosphate + NADPH + H^+ → N-acetylglutamate semialdehyde + $NADP^+$catalyzed by N-acetylglutamylphosphate reductase (NAGPR);

The semi-aldehyde formed above then undergoes a transamination reaction with glutamate as the amino group donor.

(d) N-acetylglutamate semialdehyde + Glutamate → N-acetylornithine + alpha-ketoglutarate; catalyzed by N-acetylornithine aminotransferase (NAOAT);

Ornithine can now be synthesized in two different reactions.

(e) N-acetylornithine formed as above can be converted again into N-acetylglutamate to start the cycle again. This is referred to as cyclic pathway. Along with N-acetylglutamate, ornithine also is formed in the reaction catalyzed by N-acetylornithine-glutamate acetyl transferase (NAOGAT).

(f) Alternatively, N-acetylornithine can simply be hydrolyzed to ornithine and acetate-termed as a linear pathway. The enzyme catalyzing the hydrolysis is N-acetylornithine deacetylase (NAOD).

N-acetylornithine + Glutamate → N-acetylglutamate + ornithine;

N-acetylornithine + H_2O → ornithine + acetate.

(g) Ornithine formed above may be converted to arginine as in urea cycle.

(c) Regulation

NAGS is feedback inhibited by arginine both in prokaryotes and plants. Over expression of NAGS in Arabidopsis increases ornithine accumulation and resistance to salt and drought stress. Along with ornithine other amino acids proline, arginine and citrulline also accumulate under stress conditions. NAGK also is feedback inhibited by arginine. However, feedback inhibition can be reduced by the binding specific proteins called PII to NAGK. Reduction in NAGPR causes a reduction in the production of arginine. A reduction in the levels of N-acetylornithine deacetylase (NAOD) leads to developmental abnormalities in Arabidopsis like early flowering and impaired seed setting. These abnormalities are correlated with lower levels ornithine in the affected plants.

13

Regulation of Aromatic Amino Acid Biosynthesis

General Introduction

Three amino acids, Phenylalanine, Tyrosine and Tryptophan constitute the aromatic amino acids. In prokaryotes the biosynthetic pathway is of exclusive importance only to synthesise the above three amino acids. However, in plants aromatic amino acids are further utilised for the synthesis of a huge number of phenolic compounds with important roles, for example -lignin (mechanical support to the plants. Lignin is the most abundant aromatic polymer on earth), various groups of flavonoids which have diverse roles, pharmacologically active compounds like alkaloids e.g. anti-cancer, vincristine and vinblastine, IAA (Indole -3- acetic acid) (One of the five principal hormones), colouring pigments like anthocyanins, various phenolics, Flavonoids, which can act as Phytoalexins-defensive compounds against pests and pathogens. Flavonoids and isoflavonoids can also act as inducers or anti-inducers of biological nitrogen fixation by nitrogen fixing bacteria and regulate legume-bacteria symbiosis. Phenolic compounds derived from aromatic amino acids constitute a significant fraction of plant chemical composition. This metabolism is also referred to as Phenyl propanoid metabolism or Phenolic metabolism. The above are also classified as secondary compounds. But it should be kept in mind that there are many other categories of compounds which are classified as secondary compounds, for example, terpenes, antibiotics, different drugs from medicinal plants etc. The importance of the aromatic amino acid pathway is better appreciated by the fact that approximately 20 per cent of the total carbon fixed is diverted towards the cumulative synthesis of the three aromatic amino acids.

In prokaryotes the pathway occurs in soluble portion but in plants it is localised in chloroplast.

(Students may keep in mind that many biosynthetic pathways take place in chloroplast. It is advantageous as chloroplasts can supply energy, reducing power and precursors as a result of photosynthesis).

Pathway

The pathway has several reactions and up to seven reactions till the formation of chorismate are common for all the three amino acids. From chorismate there is a branching towards the three amino acids. From chorismate, phenylalanine and tyrosine is synthesized in a sequence of three reactions. However, tryptophan synthesis from chorismate requires five and six reactions in prokaryotes and plants respectively.

The very first reaction is an important controlling point more particularly in prokaryotes. The precursors for the first reaction come from glycolysis (PEP) and OPP pathway (Erythrose 4P). Reaction may be represented as:

PEP + E4P + $H_2O \rightarrow$ 3-deoxy-D arabinoheptulosonate(DAHP) + Pi, catalyzed by DAHP Synthase. In prokaryotes this enzyme is inhibited by the end products i.e. the three aromatic amino acids. In prokaryotes, in general three isoenzymes of DAHP synthase are present, each isoenzyme specific for feedback inhibition by individual amino acids. Tyrosine and tryptophan not only feedback inhibits the respective isoenzymes but also represses DAHP synthase i.e. act at genetic level and regulate the enzyme amounts. Plants also have isoenzymes but they are not feedback inhibited by aromatic amino acids. However, the enzyme is activated by tryptophan and divalent Manganese ions in some plants.

DAHP is converted to Shikimate in three steps. Shikimate is then phosphorylated to form Shikimate 3 Phopshate by shikimate kinase;

Shikimate + ATP → Shikimate -3-Phosphate + ADP.

Shikimate kinase in plants is inhibited by its product Shikimate 3- Phosphate and ADP. The next step though it is not regulated, is of importance as it is a target for an herbicide, Glyphosate (Trade name Round up).

Shikimate -3-Phosphate + PEP → 5-enol pyruvylshikimate-3-phosphate(EPSP) + Pi.

The enzyme is EPSP synthase. The herbicide is a strong competitive inhibitor of the enzyme. This enzyme can becomes resistant to herbicide if Glycine 96 is changed to alanine. Such resistant enzymes are found naturally occurring also. The next step gives chorismate which is a common precursor for all the three amino acids.

Chorismate is isomerised to prephenate by chorismate mutase and prephenate is a common precursor for both phenylalanine and tyrosine via the formation of aerogenate. Chorismate mutase exists as two isoenzymes; one is in the plastid -CM1 and another is cytosolic -CM2. For CM1 regulation is yet to be demonstrated. CM2 is inhibited by both phenylalanine and tyrosine. Interestingly, it is also activated by tryptophan and tryptophan can also reverse the inhibition caused by phenylalanine and tyrosine leading to more production of these two amino acids. Aerogenate is formed as a result of transamination reaction with prephenate catalysed by prephenate amino transferase. The reaction is as follows:

Prephenate + glutamate → Aerogenate + alpha-keto glutarate. Aerogenate so formed is converted to phenylalanine and tyrosine by (a) aerogenate dehydratase and (b) aerogenate dehydrogenase respectively as follows:

(a) aerogenate → CO_2 + H_2O + Phenylalanine; Aerogenate dehydratse is strongly feedback inhibited by its end product phenylalanine but activated by tyrosine, the end product of aerogenate dehydrogenase.

(b) aerogenate + $NADP^+ \rightarrow$ tyrosine + NADPH + H^+; Aerogenate dehydrogenase is strongly feedback inhibited by its own end product tyrosine.

In prokaryotes like *E.coli*, it is not aerogenate that serves as the precursor as in plants-but it is prephenate which acts as a precursor for phenylalanine and tyrosine. Their formation is as follows:

(a) Prephenate $\rightarrow$ Phenylpyruvate + CO_2 + H_2O catalyzed by prephenate dehydratse; This enzyme is feedback inhibited by phenylalanine. In prokaryotes chorismate mutase and prephenate dehydratse can be a bifunctional polypeptide.

Phenylpyruvate + Glutamate $\rightarrow$ alpha-ketoglutarate + Phenylalanine.

This reaction is catalyzed by an amino transferase. The same amino transferase is also involved in the formation of tyrosine from 4-hydroxy phenyl pyruvate and is termed tyrosine amino transferase.

(b) Prephenate + $NAD^+ \rightarrow$ 4-hydroxy phenylpyruvate also called para hydroxyphenylpyruvate + NADH; Enzyme involved is prephenate dehydrogenase. It can also use $NADP^+$ and like prephenate dehydratase can also exist as a bifunctional polypeptide of chorismate mutase and prephenate dehydrogenase.

(c) 4-hydroxy phenylpyruvate so formed is transaminated to Tyrosine with glutamate as a donor of amino groups liberating alpha-ketoglutarate as shown above.

The synthesis of tryptophan from chorismate involves 5 and 6 reactions in prokaryotes and plants respectively. The first step from chorismate forms anthranilate by the action of anthranilate synthase by a transamination reaction;

Chorismate + Glutamine $\rightarrow$ Glutamic acid + pyruvate + anthranilate.

This enzyme is strongly feedback inhibited by the ultimate end product, tryptophan. Anthranilate reacts with PRPP (phospho ribosyl pyrophosphate) to form 5-phosphoribosyl anthranilate and PPi.

This reaction is catalyzed by the enzyme anthranilate phosphoribosyl transferase. This enzyme also is feedback inhibited by tryptophan.

Tryptophan is synthesized in prokaryotes as follows:

Indole 3-glycerol phosphate + serine → tryptophan + GA3P by tryptophan synthase.

In eukaryotes the enzyme has two sub-units alpha and beta which carry out two partial reactions ultimately leading to the formation of tryptophan.

(i) Indole 3-glycerol phosphate → Indole + GA3P. This reaction is catalyzed by the alpha sub-unit.

(ii) Indole + serine → tryptophan + H_2O; This reaction is catalyzed by beta sub-unit.

In prokaryotes tryptophan biosynthesis is also controlled at genetic level (regulating enzyme amounts) and by attenuation (discussed under genetic regulation).

In some organisms tryptophan biosynthetic enzymes exist as multifunctional polypeptides while in others they may exist as separate polypeptides.

As chloroplasts are semi-autonomous organelles and do not code for all the proteins required by it they have to depend on nuclear coded proteins being transported to chloroplast. Of the various enzymes involved, DAHP synthase, shikimate kinase, EPSP synthase. Phosphoribosyl anthranilate isomerase(PAI) and tryptophan synthase alpha sub-unit seem to be transported into the chloroplast. Hence this can be another regulatory step.

In view of the above mentioned roles for the various compounds arising from the aromatic amino acids, their biosynthesis is also regulated by biotic and abiotic stresses in plants. Aromatic amino acids are stress inducible.

14

Regulation of Aspartate Family of Amino Acids—Aspartate, Asparagine, Methionine, Lysine, Threonine (Cysteine, Isoleucine)

Isoleucine as a continuation of threonine metabolism and cysteine as a continuation of methionine metabolism may be discussed under aspartate family. However, keeping in view of differential classification of amino acids isoleucine is discussed under branched chain amino acids-also referred to as pyruvate family of amino acids-in chapter 15 and cysteine under serine family in chapter 16).

Aspartate

Aspartate is a non-essential amino acid but is a precursor for four essential amino acids-methionine, lysine, threonine and isoleucine- As these are essential amino acids their synthesis does not occur in mammals but only in plants and microorganisms.

Aspartate can be formed from oxaloacetae by transamination-

Glutamate + oxaloacetate → aspartate + alpha-ketoglutarate. Oxaloacetate can be formed by the carboxylation of pyruvate and PEP(discussed elsewhere). Aspartate formation, Transaminase GOT and regulation is discussed under Chapter 12. Further aspartate levels are also regulated by the activity of asparaginase which hydrolyses asparagine into aspartate and ammonia.

Asparagine

Asparagine is synthesised from aspartate by the enzyme asparagine synthetase according to the reaction:

Aspartate + ATP + Glutamine → asparagine + AMP + PPi + Glutamate with the intermediate formation of beta-aspartyl AMP.

Asparagine synthetase is important particularly in the nitrogen mobilisation during seed germination, leaf senescence and nitrogen transport from roots. In plants its activity is responsive to biotic and abiotic stresses. Over expression of asparagine synthetase increases resistance to pathogen infection. Sucrose represses the expression of asparagine synthetase. This inhibition can be relieved by supplying asparagine and other amino acids glutamate and glutamine. Since aspartate can be diverted towards the biosynthesis of aspartate family amino acids and glutamine utilised in several other reactions, it is essential that both these amino acids supply must be ensured for the asparagine synthetase reaction. Therefore, the enzymes aspartate kinase, glutamine synthetase and asparagine synthetase have to be coordinately regulated. Further the levels of asparagine are also regulated by the activity of its degradative enzyme asparaginase which hydrolyses asparagine into aspartate and ammonia. In biomedical field it is an efficient target for the treatment of several cancers.

In different organisms it can be regulated at different levels *viz.* modification of chromatin structure caused by modification of histones, transcription factors, isoenzymic control and regulation by different metabolites.

Lysine, Methionine and Threonine

Like aromatic acid biosynthesis aspartate family biosynthetic pathway also is a branched pathway and is controlled by feedback inhibition, repression and isoenzymic control.

Both in *E.coli* and plants the very first step is an important regulatory step.

(a) In the first step aspartate is phosphorylated to aspartyl-beta-phosphate also called, aspartyl-4-phosphate by aspartate kinase; Aspartate + ATP → aspartyl-beta-phosphate + ADP. There are three isoenzymes of aspartate kinase each specifically inhibited by a different amino acid to different extent. Existence of multiple forms (isoenzymes) has an advantage in metabolic pathways producing different end products. Depending on the needs of the organisms for different end products either one or more or all isoenzymes can be inhibited. Only when all the end products are not required there is a total inhibition of the reaction as all the isoenzymes are inhibited by their respective feedback inhibitors. The three isoenzymes for the above reaction may be designated as A1, A2 and A3 in microorganisms and AK1 and AK2 in plants. (However, there is no fixed rule for designation. It has been quoted so, as available in the standard text books).

A1 is allosterically inhibited by lysine and isoleucine. A2 is repressed (act at genetic level regulating the enzyme amounts) by methionine but not inhibited by any end product of the pathway. A3 is inhibited by threonine. In plants there can be either only one form of AK or two forms when they will be designated as AK1 and AK2. Where one form is present the same is inhibited by threonine, lysine and S-adenosyl methionine (SAM). Where two forms are present one form AK1 is inhibited by lysine and another form AK2 is inhibited by threonine.

(b) The next reaction provides the common precursor for all other amino acids. The precursor is aspartate beta-semialdehyde which is formed as follows:

> aspartyl-beta-phosphate + NADPH + H^+ $\leftrightarrow$ aspartate beta-semialdehyde + $NADP^+$ + Pi. The reaction is catalyzed by aspartate semialdehyde dehydrogenase (ASDH). From this intermediate one branch goes towards the biosynthesis of lysine and another branch towards the formation of homoserine. Homoserine is a common precursor for both methionine and threonine. From threonine isoleucine is formed.

(c) Homoserine is formed by the action of homoserine dehydrogenase (HSDH) from aspartate beta-semialdehyde: aspartate beta-semialdehyde + NADPH +H^+ $\rightarrow$ homoserine + $NADP^+$; In prokaryotes two isoforms of HSDH are present and are differentially inhibited. To differentially form aspartate kinase isoenzyme and avoid any confusion we may designate it differently as B1 and B2 forms of

HSDH. B1 is not inhibited by any end product but is repressed by methionine—so control at the genetic level. B2 is allosterically inhibited by threonine and isoleucine. In plants only one form is reported which is inhibited by threonine. Homoserine by successive reactions of homoserine kinase and threonine synthase forms threonine.

Homoserine + ATP $\rightarrow$ homoserine phosphate + ADP followed by homoserine phosphate + H_2O $\rightarrow$ threonine + Pi; This reaction catalyzed by Threonine synthase which catalyzes this reaction is strongly stimulated by SAM in plants.

(d) Methionine is formed from homoserine in four steps. The first step from homoserine towards methionine biosynthesis is formation of O-succinyl homoserine catalyzed by succinyl transferase also called as homoserine acyl transferase; homoserine + succinyl CoA$\rightarrow$O-succinyl homoserine + CoA. This enzyme exists in two forms. One form inhibited by methionine and another isoenzyme inhibited by threonine. Instead of succinyl CoA it can also use acetyl CoA and form O-acetyl homoserine. In the second reaction, O-succinyl homoserine reacts with cysteine to form cystathionine and succinic acid by the enzyme cysteine gamma synthase (CGS). In plants the same compound is formed from homoserine-4-phosphate and cysteine by CGS. In this reaction the additional product is Pi. CGS amounts are regulated by methionine apparently acting at genetic level.

(e) Threonine is converted to isoleucine by a series of 6 steps. The very first step is the regulatory step. Threonine is non-oxidatively deaminated to form alpha-ketobutyrate and ammonia by the enzyme threonine dehydratase. This enzyme is found in two forms in prokaryotes. One isoenzyme inhibited by isoleucine and another is repressed by high levels of isoleucine but does not exert feedback inhibition. From alpha-ketobutyrate isoleucine is formed in a series of five steps.

(f) The initial common intermediate aspartate beta-semialdehyde is used for the biosynthesis of lysine and forms lysine in a series of seven steps. The very first step is an important regulatory step: aspartate beta-semialdehyde + pyruvate $\rightarrow$ 2,3-dihydrodipicolinate + $2H_2O$.

The reaction is catalyzed by dihydrodipicolinate synthase (DHDPS). This enzyme is strongly inhibited by lysine. In both prokaryotes and plants lysine is finally formed as a result of decarboxylation of diaminopimelate by the enzyme diaminopimelate decarboxylase (DAP decarboxylase).

Isoleucine and Cysteine

As mentioned above though they can be discussed under aspartate family of amino acids they are discussed under branched chain amino acids and serine family. While isoleucine is a nutritionally essential amino acid cysteine is not.

15

Branched Chain/Pyruvate Family of Amino Acids—Isoleucine, Leucine, Valine, Alanine (Alanine which is not a Branched Chain Amino Acids is also Synthesized from Pyruvate)

General Introduction

Isoleucine, leucine and valine constitute the three branched chain amino acids. All three are nutritionally essential amino acids. Pyruvate is the carbon precursor and hence they are also called pyruvate family of amino acids. While discussing aspartate family of amino acids some mention of biosynthesis and regulation of isoleucine from threonine has been discussed. It is described here in all details.

Pathway and Regulation

(a) The first step as mentioned earlier is the non-oxidative deamination of threonine to give alpha-keto or 2-keto butyrate catalysed by the enzyme threonine dehydratase. This first step is feedback inhibited by isoleucine. (b) In the second step alpha-keto or 2-keto-butyrate reacts with hydroxyethyl-thiamine pyrophosphate (TPP) to give 2-acetohydroxybutyrate and TPP. The reaction is catalysed by the enzyme alpha-acetohydroxy synthase. This enzyme is feedback inhibited by both valine and leucine. The same enzyme also catalyses the reaction of pyruvate with hydroxyethyl-thiamine pyrophosphate (TPP) to give 2-acetolactate and TPP. 2-acetolactate is a common intermediate for the synthesis of both valine and leucine. The enzyme alpha-acetohydroxy synthase in addition to its importance as a regulatory enzyme is also important as a target for three herbicides. These herbicides are: imadazolinine, sulfonyl urea and triazolo pyrimidine. (In aromatic amino acid biosynthesis also we came across EPSP synthase as a target for the herbicide glyphosate. The practical application is as follows. We want to develop crops to possess herbicide resistant enzymes so that we can spray a higher concentration of herbicides to kill effectively only the weeds but want the crops to remain unaffected). (c) In three more steps 2-acetolactate is converted to valine via the formation of 2,3-hydroxyisovalerate and 2-ketoisovalerate. In a different direction 2-ketoisovalerate branches out to give leucine in a series of four steps.

2-ketoisovalerate reacts with acetyl CoA and a molecule of water to give 3-carboxy-3-hydroxyisocaproate by the action of 2-isopropylmalate synthase;

2-ketoisovalerate + acetylCoA + $H_2O \rightarrow$ 3-carboxy-3-hydroxyiso-caproate + CoA.

Another enzyme isopropylmalate isomerase, isomerises 3-carboxy-3-hydroxyisocaproate into 3-carboxy-2-hydroxyisocaproate which is further metabolised to leucine. This enzyme is feedback inhibited by leucine.

The final step in the biosynthesis of the above amino acid is a transamination reaction of the corresponding keto acid with glutamate as amino group donor. Glutamate after donating its amino group becomes alpha-ketoglutarate. Thus, the keto acids, alpha-ketoisovalerate, alpha-ketoisocaproate and alpha-keto-beta-methylvalerate on transamination give valine, leucine and isoleucine respectively. There may be an isoenzymic control and the number of isoforms of transaminases may be different in different organisms.

Alanine

Alanine is an unbranched amino acid and nutritionally non-essential. It is synthesized from pyruvate by transamination reaction. This transaminase GPT or alanine transaminase has been discussed earlier in Chapter 12, under the formation of glutamate.

16

Regulation of Serine Family Amino Acids- (Serine, Glycine and Cysteine)

Importance of serine, glycine and one carbon units (THF and 5, 10, methylene THF)

Serine and glycine are nutritionally not essential but otherwise indispensable. Like all other amino acids, they are required for protein synthesis. Further it is required for the synthesis of purines, histidine, thymine and methionine. Serine by 3PGA pathway is an important source in non-photosynthetic parts of the plant. The one carbon units are utilised in many metabolic reactions. Hence the biosynthesis of the above two amino acids is interconnected with many other pathways.

Serine/Glycine biosynthetic pathways

Serine

It has more than one biosynthetic pathway and consequently glycine from serine also can be formed. It is formed in plants during photorespiration and also can be synthesised from 3-phosphoglycerate. In several microorganisms it can also be synthesised from threonine.

(a) 3-phosphoglyerate pathway- From this glycolytic intermediate serine can be formed in a series of three reactions:

(a) 3PGA + NAD^+ → NADH + H^+ + 3-phosphohydroxypyruvate. This reaction is catalysed by 3-phosphoglycerate dehydrogenase.

(b) 3-phosphohydroxypyruvate then undergoes a transamination reaction with Glutamate forming alpha-ketoglutarate and 3-phosphoserine. The enzyme 3-phosphoamino transferase is involved.

(c) The final step is catalysed by phosphoserine phosphatase which hydrolyses 3-phosphoserine to serine and Pi.

(d) Serine formed as above can be converted to glycine by serine hydroxymethyl transferase (SHMT): Serine + THF (tetrahydrofolate) ↔ Glycine + 5,10, methylene THF;

(e) Glycine cleavage system/multienzyme complex (made up of four proteins) can oxidatively decarboxylate and deaminate and regenerate 5,10, methylene THF; This enzyme in eukaryotes is a mitochondrial enzyme.

Glycine + THF + NAD^+ → 5,10, methylene THF + CO_2 + NH_3 + NADH + H^+;

The 5,10, methylene THF formed in this reaction can react with another glycine molecule and synthesis serine by reversal of SHMT reaction mentioned above.

(b) Serine and glycine from threonine can be formed in a series of four reactions:

(a) threonine + NAD^+ → alpha-amino- beta-ketobutyrate + NADH + H^+-a reaction catalysed by threonine dehydrogenase.

(b) In the second reaction glycine is formed as follows by the enzyme alpha-amino- beta-ketobutyrate lyase; alpha-amino-beta-ketobutyrate + CoA → acetyl CoA + Glycine.

(c) Glycine by the above mentioned two enzymes -Glycine cleavage system and SHMT -can synthesis serine in two reactions.

(d) In photorespiration serine/glycine biosynthesis is compartmentalised. Hence a regulation by compartmentalisation exists. Glycolate formed in chloroplasts is transported into peroxisomes and gets oxidized into glyoxylate by glycolate oxidase; Glycolate + $O_2 \rightarrow$ glyoxylate + H_2O_2; H_2O_2 being toxic is detoxified by peroxisomal catalase into water and oxygen. Glyoxylate can then undergo transamination receiving an amino group either from serine or glutamate. Accordingly, the enzymes are named serine-glyoxylate transaminase and glutamate-glyoxylate transaminase:

(a) Serine + glyoxylate $\leftrightarrow$ Glycine + hydroxypyruvate;

(b) Glutamate + Glyoxylate $\leftrightarrow$ alpha-ketoglutarate + Glycine.

Glycine formed as above can be acted upon by the above mentioned enzymes Glycine cleavage system and SHMT in mitochondria to give serine which can be transported back to peroxisome. The glycine cleavage reaction is the source of photorespiratory CO_2 which is released. Ammonia simultaneously released in the reaction is re-assimilated by the chloroplasts.

Regulation of Serine/Glycine biosynthetic pathways

3PGDH has three isoforms and therefore is subjected to isoenzymic control. This enzyme is allosterically inhibited by serine, alanine and glycine and to some extent by alpha-ketoglutarate as well. 3-phosphoamino transferase has two isoforms. Purines can repress Glycine cleavage system. The multienzyme is also inhibited by post-translational modification of its cysteine residues either by S-nitrosylation or S-glutathionylation. Nitric oxide also can inhibit this enzyme. Availability of threonine is another factor that can regulate the metabolic route chosen for the biosynthesis of serine. If more amount of threonine is available, then serine biosynthesis preferentially takes place starting with threonine. Otherwise 3PGA pathway is preferred. SHMT and Glycine cleavage system are regulated at operon level in microorganisms.

Cysteine

(a) General Introduction

Cysteine is nutritionally non-essential. Cysteine also has more than one biosynthetic pathway. Mammals can synthesise it from methionine and hence it is not considered as nutritionally essential. Plants and bacteria can synthesize cysteine from serine. Hence cysteine can be considered as belonging both to serine family and also aspartate family in which it is a further metabolism of methionine formed.

(b) Cysteine Biosynthetic pathway from Methionine and Regulation

(a) Pathway

(i) Methionine + ATP SAM (S-adenosylmethionine) + Pi + PPi. Reaction catalyzed by methionine adenosyl transferase (MAT).

(ii) In the second reaction SAM donates its methyl group for so many acceptors.

SAM + acceptor of methyl groups $\rightarrow$ CH_3-acceptor + adenosine + L-Homocysteine. During this reaction S-adenosylhomoysteine is formed as an intermediate which gets hydrolysed to adenosine + L-Homocysteine. The reaction is catalysed by a methyl transferase with a broad specificity for acceptors.

(iii) Serine + Homocysteine $\leftrightarrow$ H_2O + L- cystathionine. It is a transsulfurylation reaction catalysed by cystathionine beta synthase (CBS). Cystathionine is hydrolysed to give L-cysteine and L-homoserine by cystathionine gamma lyase (CGL) also called cystathionase.

L- cystathionine + H_2O $\rightarrow$ L-cysteine +NH_3 + alpha-ketobutyrate;

(iv) In some organisms L-cystathionine can react with succinate to give L-cysteine and O-succinylhomoserine which is a reversible reaction of Cystathionine-gamma-synthase. In other words, in the other direction,

(v) L-cystathionine and succinate are formed from O-sucinylhomoserine and L-cysteine.

O-succinylhomoserine is formed from succinyl CoA and L-homoserine with the simultaneous liberation of CoA by the enzyme homoserine O-succinyltransferase.

(b) Regulation

From the above pathway it is evident that the sulfur of cysteine is derived from methionine while the carbon skeleton is from serine. The above two reactions are very important. SAM donates methyl group to several important acceptors and thus is interconnected with so many pathways. Two examples worth mentioning is methylation of DNA and histones which can regulate transcription of so many genes accompanied or unaccompanied by chromatin reorganisation. Hence methyl transferases can be regulated by so many metabolites and at so many levels for example, regulation at post-transcriptional and post-translational level has been demonstrated in case of DNA methyl transferases which also uses SAM as a methyl group donor.

MAT has isoenzymic control and also regulated by oligomerisation status. Differential regulation of isoforms of MAT are implicated in cancer progression.

Cystathionine beta synthase (CBS) is a multienzyme complex and is pyridoxal phosphate dependent. (In some organisms the enzyme also has heme in which case the valency status of iron in heme may regulate the activity. Ferric form Fe^{3+} activates the enzyme). It is activated by SAM. In some microorganism it can also be inhibited by cysteine and disulfide and the levels of cysteine are also dependent on the rate of its degradation by cysteine desulfhydrases. Cysteine operon is both negatively and positively regulated. These two can also inhibit. It is also subjected to hormonal regulation at transcriptional level mediated by specific transcription factors. The enzyme is upregulated by glucocorticoids and downregulated by insulin. Methionine is implicated in upregulation at post-transcriptional level. There is over expression of Cystathionine beta synthase in case of the abnormal medical condition in children -Down's syndrome.

Hydrogen sulfide depending on its concentration can either decrease, increase or completely inhibit transcription of cystathionine gamma lyase (CGL). The enzyme is sensitive to free radicals and is implicated in several diseases like cancer, AIDS, Diabetes, blood pressure, cystathionuria, etc. On the other hand, hydrogen peroxide can upregulate the expression of both mRNA and protein of cystathionine gamma lyase (CGL).

(c) Cysteine biosynthetic pathway from serine and its regulation

Serine + acetyl CoA → O-aetylserine + CoA catalysed by serine acetyl transferase.

-L-cysteine is formed in the second reaction as follows:

O-actylserine + disulfide S^{2-} → L-cysteine + acetate (CH_3COO- by the enzyme

O-actylserine (thiol) lyase. Sulfate in plants and microorganisms can be reduced to disulfide.

Cysteine inhibits serine acetyl transferase and represses O-actylserine (thiol) lyase. Increased expression of the above two enzymes by molecular approaches in Arabidopsis improves the plants resistance to heavy metals and protection against oxidative stress. In Arabidopsis isoenzymic control also is there as there are three isoforms. Cysteine in all organisms is an important constituent of glutathione which is important in antioxidant defense. In plants specifically it is important for the synthesis of metal binding peptides called phytochelatins. It is also required for several secondary metabolites.

17

HISTIDINE

General Introduction

(a) General Introduction

It is nutritionally an essential amino acid. Histidine is required for some secondary plant metabolites. The intermediates in the histidine biosynthetic pathway—imidazole glycerol phosphate and amino-imidazole-4-carboxamide ribonucleotide (AICAR) are interconnected with purine biosynthetic pathway. It is of importance in several physiological processes in humans. Histidine is essential for zinc absorption and transport to tissues.

Histamine the decarboxylated metabolite of histidine play an important role as neurotransmitter, immunomodulation, inflammatory responses and sexual functions both in men and women. Histidine biosynthesis is pathway present in bacteria, archaea, lower eukaryotes, and plants. It is a linear pathway with ten enzymatic reactions, starting with phosphoribosyl pyrophosphate (PRPP) and leading to histidine. In microorganisms usually a polycistronic operon controls the transcription of the genes i.e. all genes are transcribed as a single unit. However, in other microorganisms and other organisms the genes can be scattered along the genome.

(b) Pathway and Regulation - The very first step in all organisms

PRPP + ATP $\rightarrow$ 1-(5-phospho-D-ribosyl)-ATP also written as 5'-phosphoribosyl-ATP (PRATP) + PPi is catalysed by ATP phosphoribosyltransferase. It is the most important enzyme being regulated. Regulation at this step avoids expenditure of energy in the form of ATP and PRPP (Recollect that PRPP is derived from R5P from OPP pathway). Wastage of PRPP needs to be avoided as it is also involved in the *de novo* synthesis of purines and pyrimidines, tryptophan biosynthesis in all organisms, and for the synthesis of arabinogalactan, an important component of the Corynebacterium cell wall. *Corynebacterium glutamicum* is the microorganism of the choice for the large scale production of glutamate and lysine and being exploited for the large scale production of several other amino acids. Further the role of intermediates in the histidine biosynthetic pathway -imidazole glycerol phosphate and aminoimidazole-4-carboxamide ribonucleotide (AICAR) in purine biosynthesis enhances the importance of this regulation. This enzyme is allosterically inhibited by the final end product histidine. Transcription of the genes can be regulated by stringent response. The accumulation of guanosine tetraphosphate (ppGpp) and guanosine pentaphosphate (pppGpp) transcription is usually upregulated. Histidine biosynthesis is also regulated by attenuation. The inhibitory effect of histidine can be enhanced in case of prokaryotes by ppGpp. The last two steps which converts histidinol to histidine are catalyse by histidinol dehydrogenase which is also feedback inhibited by histidine:

$$\text{Histidinol} + NAD^+ \leftrightarrow \text{histidinal} + NADH + H^+;$$

$$\text{Histidinal} + NAD^+ + H_2O \rightarrow \text{histidine} + NADH + H^+.$$

18

Strategies for Over Production of Amino Acids and other Metabolites

General Introduction

The study of regulation of amino acid metabolic pathways and the strategies employed by scientists for overproducing the desired amino acids should give the students an understanding of how different metabolic pathways -even other than amino acid biosynthesis- may be manipulated for obtaining desired products. After ethanol and antibiotics amino acids constitute the major commercial products in microbial biotechnology. All the 20 amino acids find application in some industry or other. For example, food and feed additives, therapeutics, flavour enhancers, cosmetics, etc. As mammals cannot synthesise essential amino acids there are attempts to produce plants whose seeds contain more amount of essential amino acids. These strategies would be useful to all categories of biotechnologists both using conventional approaches and as well as advanced molecular techniques. Though many or even most are employing advanced molecular techniques to achieve the goals. But their understanding of how to manipulate is derived from the understanding of the metabolic regulation and metabolic pathways which have been discussed. The strategies may be summarised as follows: (Employing combination of all these strategies to achieve the desired goal is called metabolic engineering). While discussing metabolic pathways other than that of amino acids many manipulations involving molecular techniques have been mentioned. To have a better understanding of all the above, the section on genetic regulation and signal transduction may be studied.

Strategies

(a) If the system is a microorganism, improved strains may be obtained by classical strain improvement techniques like mutagenesis, selection etc. Alternatively, a suitable microorganism with less complex regulation may be selected. Useful mutants can be naturally occurring. Say for example, organisms/varieties/strains with regulatory enzymes insensitive to feedback inhibition may be found in nature. Either the above or genes for such enzymes can be isolated and cloned and expressed into the organism chosen for achieving the goal.

(b) Use of antimetabolites or analogues of endogenous feedback inhibitors. For example, 5-fluro or chloro phenylalanine and 5-methyl tryptophan are analogues of phenylalanine and tryptophan respectively. Histidine analogues are; 2-methyl histidine and 1,2, 4-triazole-3-alanine. We have analogues for other amino acids too. These bind to the allosteric site of the concerned enzyme/isoenzyme in the biosynthetic pathway, which is otherwise inhibited by the end product/endogenously occurring feedback inhibitor. As a result of this treatment only strains which are resistant to antimetabolites would survive. Such strains would also be insensitive to endogenously occurring feedback inhibitors. Hence the concerned regulatory enzyme is no more subjected to feedback inhibition and would continue to function in the metabolic pathway.

(c) In a branched metabolic pathway, competing pathways –genes of those products are not required- may be knocked off by molecular techniques like gene disruptions etc. (See discussion under genetic regulation).

(d) Gene amplification i.e. creating more copies of the genes required in the selected system, which will result in the over expression of the desired enzyme and naturally more production of the metabolite by the concerned enzyme. In a branched metabolic pathway amplification of selected genes is required both in the common metabolic pathway for the production of many metabolites and in the branched pathway where only a particular product is produced. (However, a point which students and molecular scientists should remember is over expression of genes is possible only if the primary metabolic pathways are continuously operating. Say a scientist has over expressed a gene. It means, first it results in mRNA abundance i.e. more number of mRNA molecules. What is mRNA? A polynucleotide of four nucleotides and the RNA polymerase requires these four nucleotides in the form of NTP's, ATP, UTP, CTP and GTP. So over expression of a gene resulting in abundance of mRNA's requires increased synthesis of purines and pyrimidines. So the purine and pyrimidine biosynthetic pathways have to be activated. And the synthesis of purines and pyrimidines requires several precursors/processes *viz.* aspartate, glycine and glutamine and one carbon metabolism which supplies formyl-tetrahydrofolate for purine synthesis to continue. The carbon skeletons for amino acids are supplied from glycolysis and TCA cycle. These cycles also supply energy in the form of ATP. The pentose sugar is provided by the OPP pathway. Same holds good for pyrimidine biosynthesis. Their precursors are aspartate and glutamine apart from bicarbonate formed from CO_2. Further for the proteins to be synthesised from mRNA's all the 20 amino acids have to be synthesised and supplied in required amounts corresponding to the polypeptide composition (See amino acid metabolism).

(e) Removing control points at all levels whichever is/are relevant i.e. repression, attenuation and feedback inhibition. This can be achieved by eliminating or inactivating repressor genes, removal or alteration of operator and/or attenuator sequences (in case the product is regulated by attenuation) and the use of genes encoding feedback resistant variations of rate limiting or regulatory enzymes.

(f) Increase precursor(s) supply both in common metabolic pathway and as well as in the specific branch.

(g) The desired product, though formed may be catabolised. Hence degrading enzyme(s) have to be inactivated or their synthesis stopped.

Details of some of the strategies employed

The discussion below explains how some of the above mentioned strategies have been employed in the biosynthetic pathway of certain amino acids and the problems that may be faced. This should give an idea for manipulating other metabolic pathways also. Strategies can be applied not only to microorganisms but also in tissue culture technology. Further useful mutants of intact plants may be obtained to achieve the desired goal. There can be naturally occurring useful mutants which may be exploited.

(a) Strategies to increase precursor supply

Say in aromatic amino acid pathway the precursors are PEP and E4P. Their supply need to be increased. (a) For increasing PEP its diversion to other metabolic reactions i.e. competing reactions have to be removed either by inactivation *viz.* PEP carboxylase and Pyruvate kinase. Then more PEP will be diverted to AAA pathway. (b) PEP precursor supply can also be increased by activation of PEP carboxy kinase which produces PEP from oxaloacetate (oxaloacetate → PEP). (c) Cloning and overproduction of TK and TA in OPP pathway to increase E4P supply. (d) In some cases a microorganism may take up sugars from the medium using an energy source which could be PEP. Hence such uptake must be minimised or alternatively organisms/strains which can take up sugars otherwise without depending on PEP should be selected. In case of tryptophan in addition to above precursors two more precursors also are required; Serine and PRPP. Hence their supply needs to be increased by increasing serine biosynthetic pathway and over production of PRPP synthetase. (However, PRPP is used up not only in tryptophan biosynthesis but also diverted towards nucleotide biosynthesis). (as metabolism/metabolic pathways are is interconnected diversion of precursors towards the desired product may cause some undesirable side effects on the organism making it unsuitable for the desired goal. Hence there should be a balanced supply of precursors).

(b) Eliminating competing pathways

Say if out of the three aromatic amino acids only overproduction of tryptophan is required, then branches leading to the synthesis of other two amino acids phenylalanine and tyrosine have to be controlled.

(c) Inactivation of degrading enzymes

Proline degrading enzyme prolyl oxidase has been inactivated to ensure over accumulation of proline. Similarly, tryptophan degrading enzyme, tryptophanase is inactivated. Lysine-an essential amino acid-may be over produced by over expression or by having a feedback insensitive biosynthetic enzyme like dihydrodipicolinate synthase. However, the overproduction may be nullified by its catabolism. It is catabolised by two enzymes, lysine alpha-ketoglutarate reductase which converts lysine to saccharopine;

Lysine + NADPH+ $H^+ \rightarrow$ Saccharopine + $NADP^+$; Saccharopine is in turn catabolised to alpha-aminoadipic semi-aldehyde + glutamate by saccharopine dehydrogenase. In the process NAD^+ is reduced to NADH + H^+.

(d) Over expression of biosynthetic enzymes

Proline finds an application not only in industry but plays a major role in plants in protecting them from water stress caused by drought or salinity or temperature. It is one of the important osmolytes accumulated during abiotic stresses as above. Over expression of the biosynthetic enzyme, pyrroline-5-carboxylate synthetase (PC5 synthetase) caused a 14 fold increase in proline accumulation in moth bean (*Vigna aconitifolia*) and made the crop more resistant to drought and salinity. Since water stress can cause an increase in proline accumulation, this strategy, that is subjecting the microorganism to osmotic stress for further increasing the proline accumulation has been employed in microorganisms in addition to the other strategies to increase its production. (However, students should remember that proline is not the only osmolyte/osmoprotectant that is accumulated under abiotic stress. Different plants may differ in the type of osmolyte accumulated. In addition to proline, some important osmolytes are glycine betaine, mannitol, sorbitol, ononitol, trehalose etc. Hence strategies mentioned above have to be accordingly followed. That is the biosynthetic pathways of these osmoprotectants must be studied and regulatory steps identified and isolate the relevant genes either from the plant concerned or from microorganisms. These genes have to be cloned into the relevant plant which accumulate the particular osmolyte. In addition to the metabolites stress resistant proteins also are found called osmotins. Over expression of these genes may confer stress resistance.

(e) Accumulation of the desired product within the cell can be deleterious. Further if the desired product is accumulated within the cell the cost of the recovery of the desired product from the cell will increase. Hence it should be facilitated that the desired products are excreted from the cells into the medium, so that the recovery of the product is easier and less expensive—referred to as downstream processing costs. If there are specific excretory systems for the desired products their activity should be increased.

Few Examples of Applications

A few applications achieved using the above strategies and some additional techniques are described here.

(a) Production of herbicide resistant plants

It has been mentioned earlier that EPSP synthase is one of the important reactions in aromatic amino acid biosynthesis and is a target for developing Glyphosate (Round Up) herbicide resistant plants. As we want the plants to be resistant to the herbicide there is a need to develop herbicide resistant plants. (a) one way is by over expression of the enzyme by gene amplification i.e. by introducing multiple copies of the gene and using a strong promoter to enhance transcription. The excess protein will protect the plant as enough enzyme is available for amino acid biosynthesis even after some amount of enzyme is inhibited by the herbicide glyphosate, (b) The basis for glyphosate resistance lies in a change of amino acid of EPSP synthase- substituting an alanine residue for a glycine residue at position 96. (other positions also may be substituted to develop resistance where in either an aspartic acid residue or asparagine residue is substituted for glycine residue wild type EPSP synthase) Hence such mutated proteins may be cloned into plants. Alternatively, EPSP synthase resistant or less insensitive to herbicide may be found in other organisms, for example, in some Agrobacterium strains and salmonella. Such genes may be cloned into plants, (c) another strategy is to detoxify the herbicide by acetylation or oxidation. Detoxifying enzymes, acetyl transferase and glyphosate oxidase are available in certain microorganisms.

Using the above strategies herbicide resistant varieties have been developed in maize, soybean and rapeseed. For other herbicides similar strategies may be employed.

(b) The development of Genetically modified crops towards lepidopteran pests by cloning cry genes which codes for cry protein toxins is based not on the knowledge of biosynthetic pathways in plants but on the mode of action of the toxin on the insect. The toxins kill the insects when they feed on the plants.

(c) Production of plants with increased content of some essential amino acids in the seed proteins.

(a) genes for methionine poor proteins may be isolated and introduce mutations to increase codons for methionine. Such genes may be cloned into plants, (b) already existing methionine rich protein genes available from different crops may be used to clone into other plants, (c) over expression of cystathionine -gamma-synthase, the first enzyme in methionine biosynthesis in alfalfa increased in proteins content rich in methionine. (However, as mentioned above the required supply of the amino acid may not be assured. Its biosynthetic activity needs to be increased. Hence there can be only limited or no success with such approaches). Similar approaches have also been tried for lysine too.

19

UREA CYCLE

General Introduction

Urea Cycle also called Krebs-Henseleit cycle. Different organisms have adopted different strategies to dispose/excrete their nitrogenous waste. Accordingly, they are classified as (a) ammonotelic or ammonotelism -e.g. aquatic animals and many invertebrates. They excrete nitrogen in the form of ammonia. (b) Uricotelic or uricotelism- e.g. insects, birds many reptiles like snakes, lizards etc. they excrete nitrogen in the form of uric acid. (c) Ureotelic or ureotelism e.g. mammals, many adult amphibians etc. They excrete nitrogen in the form of urea.

Nitrogenous waste arises as a result of protein breakdown/turnover. The resulting amino acids are further metabolised by different deamination reactions which results in ammonia which is toxic and has to be excreted. As a result of protein turnover particularly the amino acids glutamine, glutamic acid enters mitochondria. Other amino acids which undergo deamination reactions in cytosol can be incorporated in to glutamic acid and glutamine by the enzymes glutamate dehydrogenase (GDH) and glutamine synthetase (GS). GDH catalyze the reaction-alpha-KG + NADPH $+H^+ + NH_4^+ \leftrightarrow$ Glutamate + $NADP^+$ + H_2O. GDH is a reversible reaction. Hence either ammonia can be liberated or glutamate can be formed.

GS catalyzes the reaction- Glutamate + NH_4^+ + ATP $\rightarrow$ Glutamine + ADP + Pi.The enzyme requires divalent magnesium ions. The glutamate and glutamine arising as above can liberate ammonia in mitochondria by the action of mitochondrial GDH and glutaminase. While protein turnover is a normal continuous process, there can be special situations where ammonia formation and consequently urea formation is increased. Starvation leads to excess protein breakdown and amino acid metabolism resulting in more ammonia. Similarly, any excess intake of Protein/amino acids can lead to formation of ammonia. The other waste product that participates in the formation of urea is CO_2 resulting from respiration and decarboxylation reactions.

Biosynthesis

Liver is the primary site for the biosynthesis of urea. To a small extent the process also occurs in kidney. Finally, urea is excreted through kidneys in the urine.

Urea cycle is compartmentalised between mitochondria and cytosol and is linked to TCA cycle and gluconeogenesis. As energy is required for urea formation it has to come from glycolysis/TCA cycle. Further, many intermediates of urea cycle are also intermediates of TCA cycle. Out of the five enzymes involved, only two-carbamoyl phosphate synthetase-1 (CPS-1) and ornithine transcarbamoylase (OTC) are present in mitochondria. The other three enzymes, arginino succinate synthase (ASS), arginino succinate lyase (ASL) - (also called arginino succinase) and arginase are present in cytosol; CPS I is so designated to distinguish it from another CPS involved in pyrimidine biosynthesis in cytosol which uses glutamine rather than ammonia. This is designated as CPS II. CPS I catalyses the reaction; $NH_4^+ + HCO_3^- + 2\,ATP \rightarrow$ Carbamoyl phosphate (CP) + 2 ADP + Pi; and ornithine transcarbamoylase (OTC) are present in mitochondria. The other three enzymes, arginino succinate synthase (ASS), arginino succinate lyase (ASL) -(also called arginino succinase) and arginase are present in cytosol.

Regulation

(a) Compartmentalisation

The reactions are distributed between mitochondria and cytosol. Hence flow of metabolites between the compartments must be regulated.

(b) CPS-I is the major regulatory step. CPS-I is allosterically activated by N-acetyl glutamate. This regulatory metabolite is synthesized by N-acetyl glutamate synthase from acetyl CoA and glutamate.

acetyl CoA + glutamate → N-acetylglutamate + CoA. So high levels of acetyl-CoA and glutamate result in the synthesis of high amounts of N-acetylglutamate.

Further N-acetyl glutamate synthase is activated by arginine (arginine on hydrolysis by arginase gives urea and ornithine).

(c) The enzyme amounts of all the above mentioned five enzymes are altered- increased- during increased protein break down during starvation or excess dietary intake of protein. In addition to the five enzymes glutamate resulting from the increased protein break down also increases glutamate and thereby increase the formation of N-acetylglutamate. This will stimulate the activity of CPS-I and consequently urea cycle. But when the dietary level of protein is very low these enzyme amounts are decreased.

(d) Interrelationship with TCA cycle

As mentioned above only the first two enzymes concerned with urea cycle are present in mitochondria. Others are in cytosol. After the formation of CP, it combines with ornithine and forms citrulline, a reaction catalysed by OTC in mitochondria. Citrulline is transported out of mitochondria into cytosol and there it combines with aspartate and ATP to give arginino succinate (AS), AMP and PPi. This reaction is catalysed by ASS. Aspartate used in this reaction comes from mitochondria and is possible to come from cytosol too. AS is cleaved by ASL to yield arginine and fumarate. (Here fumarate is formed in the cytosol. Otherwise fumarate is also an obligatory intermediate in TCA cycle). Irreversible hydrolysis of arginine by arginase results in the formation of urea and ornithine is regenerated to start the cycle again by combining with CP. So ornithine formed in the cytosol has to be transported to mitochondria. Fumarate formed in ASL reaction can be converted into malate and oxaloacetate in the cytosol by cytosolic fumarase and MDH or can be exported as such to mitochondria.

Fumarate, malate, oxaloacetate and aspartate formed in the cytosol may be diverted towards gluconeogenesis.

Urea generation in plants

Plants may get urea both exogenously and endogenously. When nitrogen fertiliser is added in the form of urea it may be taken up- to varying amounts- as such- or taken up as ammonia after urea is broken down by soil microorganisms into ammonia and CO_2. Urea so taken up is simply hydrolyzed inside the plant by urease into ammonia and CO_2. Many legumes under nodulated conditions synthesize ureides-allantoin and allantoic acid- as the end products of biological nitrogen fixation and they are transported to the aerial parts. There ureides are finally broken down to urea and then by urease into ammonia and CO_2. Ammonia so released is refixed into amino acids by the plant enzymes GS and GOGAT-glutamine oxoglutarate amidotransferase-(also called glutamate synthase). Plants also possess arginase and hence urea may be generated by the hydrolysis of arginine. Urea so formed is broken down by urease. Plant arginases are supposed to be induced by external pressures like wounding, pathogenic attack, etc.

Part - V

Regulation of Nucleic Acid Metabolism

20

Purine Biosynthesis

General Introduction

Both DNA which forms the genome in all forms of life and its primary products different RNA's are polymers of purine and pyrimidine nucleotides. Purine bases are adenine (6-aminopurine), Guanine (2-amino 6-oxypurine). While forming DNA or RNA the concerned enzymes DNA polymerase and RNA polymerase uses NTP's (Nucleoside triphosphates) as substrates and incorporate only the NMP (nucleoside monophosphate portion) liberating PPi.

Purine biosynthesis is metabolically highly interconnected. The precursors of a purine ring are (a) Amino acids -glycine, aspartate and glutamine-the carbon skeletons for these amino acids come from glycolysis/TCA cycle intermediates. Amino group of amino acids in case of plants and microorganisms can come as a result of nitrate or ammonia assimilation while in heterotrophic organisms they come from other amino acids or free ammonia. All the amino acid precursors of purines are non-essential amino acids and hence can be synthesised by the heterotrophic organisms. (b) Tetrahydrofolate derivatives from one carbon metabolism. (c) CO_2-may arise from respiration or decarboxylation reactions. (d) Of the 9 atoms of the purine ring, 3 and 4 are derived from glutamine; 2 and 8 are derived from folate derivatives; 1 from aspartate; 4,5,7 from glycine and 6 from CO_2. (e) For the purine ring to become a nucleotide it requires the pentose sugar ribose (in case of RNA's) or deoxyribose (in case of DNA) which come from OPP pathway, and a phosphate group. (f) The biosynthesis requires energy in the form of ATP which comes from substrate or oxidative level phosphorylation during the oxidation of metabolic fuels. In plants and other photosynthetic organisms energy in the form of ATP can be produced during light reactions-photosynthetic phosphorylation also.

The synthesis of a purine ring/nucleotide starting from the basic precursors is termed as *de novo* biosynthetic pathway. Starting from R5P purine synthesis consists of 13 reactions for the formation of either AMP or GMP. IMP (Inosine monophosphate) is the first purine nucleotide that is formed in the biosynthetic pathway at the end of 11 reactions. For the conversion of AMP or GMP into their corresponding triphosphates i.e. ATP and GTP two more reactions are required. A nucleotide constitutes a base, pentose sugar and a phosphate. In purine synthesis sugar phosphate is first synthesised followed by the base while in case of pyrimidine biosynthesis, base is synthesised first followed by the addition of sugar phosphate.

Regulation

Of all the reactions only five are major regulatory steps:

(a) Ribose phosphate pyrophosphate kinase also called phosphoribosyl pyrophosphate synthetase (PRPP synthetase). This enzyme catalyses the first reaction in the pathway.

R5P + ATP $\rightarrow$ PRPP + AMP. This enzyme is primarily feedback inhibited by ADP and GDP. ADP is formed from AMP which is in turn formed from IMP. Similarly, GDP is formed from GMP which in turn is formed from IMP. Hence high levels of IMP, AMP and GMP also may be said to inhibit the enzyme. Further it can also be feedback inhibited by pyrimidine nucleotides. It should be kept in mind that PRPP is used both in purine and pyrimidine synthesis. So to coordinate their synthesis PRPP synthetase has to be regulated both by purine and pyrimidine nucleotides.

(b) Amidophosphoribosyl transferase also called glutamine -PRPP amidotransferase -This enzyme catalyses the second reaction in the pathway as follows:

PRPP + glutamine + H_2O $\rightarrow$ beta-phosphoribosylamine (PRA) +

glutamate + PPi.

This enzyme is inhibited by several nucleotides *viz.* AMP, ADP and ATP, GMP, GDP and GTP, IMP and XMP (xanthosine monophosphate). This enzyme is also subjected to feed forward activation by PRPP.

(c) IMP dehydrogenase-IMP is the first purine nucleotide that is formed. Separate branches lead to the formation of AMP and GMP in two more steps.

IMP dehydrogenase catalyses the reaction-IMP+ $NAD^+ + H_2O \rightarrow$ XMP + NADH + H^+. This enzyme is inhibited by GMP.

(d) Adenylosuccinate synthetase (ASS)-By this enzyme IMP can take another route leading to the formation of AMP. It catalyses the reaction-

IMP + aspartate + GTP $\rightarrow$ Adenylosuccinate + GDP + Pi. The enzyme is inhibited by AMP and GTP. Adenylosuccinate so synthesised is converted to AMP by adenylosuccinate lyase (ASL) with the simultaneous liberation of fumarate.

(e) XMP formed in the IMP dehydrogenase reaction is converted to GMP by the action of GMP synthetase. Reaction may be depicted as:

XMP + glutamine + ATP + $H_2O \rightarrow$ GMP + glutamate + AMP + PPi. This enzyme is inhibited by ATP.

(f) Taking 'D' and 'E' together we may say that ATP inhibits GTP synthesis and GTP leads to the inhibition of ATP synthesis.

(g) AMP and GMP formed as above can be converted to their respective triphosphates at the expense of ATP by NMP (nucleoside monophosphate) and NDP(nucleoside diphosphate) kinases as follows:

(i) AMP/GMP + ATP $\rightarrow$ ADP/GDP + ADP;

(ii) ADP/GDP + ATP $\rightarrow$ ATP/GTP + ADP.

Salvage pathway

Pathways of purine synthesis i.e. AMP, GMP, IMP using preformed purine bases is called **salvage pathway**. Deficiencies in the enzymes

of salvage pathway can lead to genetic abnormalities like Lesch-Nyhan syndrome (sex linked inborn error in males) and immuno deficiency. Enzyme deficient in the above syndrome is Hypoxanthine-guanine phosphoribosyltransferase (HGPRT) which catalyses the reaction;

Hypoxanthine/Guanine + PRPP → IMP/GMP + PPi.

The enzyme deficiency leads to the accumulation of PRPP which activates amidophosphoribosyl tarnsferase (mentioned above) leading to excess purine synthesis. Excess purine synthesis in turn leads to excess purine degradation causing an accumulation of uric acid. People suffering from this syndrome have high levels of uric acid and exhibit several neurological abnormalities.

Uric acid accumulation also causes the disease Gout –a type of arthritis –causing painful joints.

Excessive uric acid production is also caused by glucose-6-phosphatase deficiency. This syndrome is referred to as von Gierke's glycogen storage disease. The drug allopurinol inhibits the enzyme xanthine oxidase responsible for uric acid formation, decreases the level of uric acid and improves the condition of the patient. However, the drug has no effect on neurological abnormalities of Lesch-Nyhan syndrome. Therefore, some other factors other than uric acid are involved in the above syndrome. Another enzyme of salvage pathway is Adenine phosphoribosyl transferase (APRT) which catalyses the reaction; Adenine + PRPP → AMP + PPi.

21

Pyrimidine Biosynthesis

General Introduction

As mentioned above pyrimidine ring is formed first in the pathway followed by the addition of sugar phosphate. Precursors for pyrimidine ring are aspartate, glutamine and bicarbonate (HCO_3^-). Of the six atoms, 1,4,5 and 6 are derived from aspartate;

2 from glutamine and 1 from bicarbonate. Pyrimidine NTP's are CTP, TTP and UTP. In pyrimidine biosynthetic pathway initially UTP and CTP are formed. UTP is formed via UMP and UDP while CTP is formed from UTP. TTP can be formed from both CDP and as well as UDP. Uracil (2,4, dioxypyrimidine) is found only in RNA and hence exists as ribose conjugate only, whereas Thymine (2,4,dioxy-5 methyl pyrimidine) found only in DNA is conjugated only with deoxyribose sugar. The other pyrimidine cytosine (2-oxy-4-aminopyrimidine) should be made available both as a ribose and deoxyribose conjugate as it is found both in RNA and DNA. Similarly, with the purines adenine and guanine. Hence there must be a mechanism to convert ribose sugar into deoxyribose.

Regulation

As mentioned earlier CPS II is involved in pyrimidine biosynthesis. While eukaryotes have CPS I and CPS II, prokaryotes have only one CPS. Both catalyse the reaction:

Glutamine + 2 ATP + bicarbonate (HCO_3^-) + H_2O
$\rightarrow$ Carbamoyl phosphate (CP) + glutamate + 2 ADP + Pi.

(a) While CPS II is regulated prokaryotic CPS is not. CPS II is activated by PRPP and ATP and inhibited by UDP and UTP. Formation of PRPP and Regulation of PRPP synthetase has been discussed under purine synthesis.

(b) In addition to CPS II only one more enzyme OMP (orotodine 5 prime monophosphate) decarboxylase is regulated in eukaryotes. OMP is the first pyrimidine nucleotide formed in the pathway which leads to the synthesis of physiologically relevant pyrimidine nucleotide UMP by decarboxylation. Reaction may be represented as:

OMP → UMP + CO_2. The enzyme is inhibited by its own product UMP.

(c) In prokaryotes the immediate reaction after the formation of CP is:

CP + aspartate → carbamoyl aspartate + Pi. This is catalysed by the enzyme aspartate transcarbamoylase and is an important step for regulation. This enzyme is activated by ATP and inhibited by CTP the final end product of the pathway-the last committed step.

(d) Orotate is converted to orotidylate (OMP) by the action of orotate phosphoribosyl transferase (OPRT); Orotate + PRPP → OMP + PPi. This enzyme is a target for the anti-cancer drug 5-flurouracil.

(e) OMP so formed is decarboxylated to form UMP by Orotidylate decarboxylase (ODC) also called orotidine-5'-monophosphate decarboxylase (OMPDC); OMP/orotidylate → UMP (uridylate) + O_2.

Allopurinol, a purine analog which inhibits purine metabolism and formation of uric acid responsible for gout disease, in addition to xanthine oxidase also inhibits Orotidylate decarboxylase(ODC) also called orotidine-5'-monophosphate decarboxylase (OMPDC) and thus influences the production of pyrimidine nucleotides as well.

(f) UMP is converted to UDP and in turn to UTP by NMP and NDP kinase respectively. (NDP kinases can act both on NDP's and also on dNDP's) UTP in turn gives rise to CTP by the action of CTP synthetase as follows:

UTP + glutamine + ATP+ H_2O → CTP + glutamate + ADP + Pi;

CTP synthetase is regulated in many ways in humans. (i) The enzyme is inhibited by its own end product CTP. (ii) The enzyme is also regulated by oligomerisation status i.e. subunit association and dissociation. The active enzyme is a homotetramer. When ATP and UTP are absent/deficient the enzyme exists as an inactive monomer. In the presence of sufficient ATP and UTP active tetrameric status is restored. This restoration is further facilitated if the enzyme is present at a higher concentration. (iii) GTP acts as an allosteric activator by stimulating the hydrolysis of glutamine, but at higher concentrations it can inhibit CTP formation. (iv) Humans have two isoenzymes CTPS I and CTPS II. Hence some isoenzymic control must be there. (v) CTPS I is inactivated by phosphorylation either at Serine 568 or Serine 571 by specific protein kinases. (vi) In humans some enzymes are regulated at genetic level by repression and derepression. Main enzymes include:

CPS II, ATC, dihydroorotase (DHO) (which dehydrates N-carbamoylaspartate to form dihydroorotate), OPRT and OMPDC.

22

Conversion of Ribonucleotides into Deoxyribonucleotides

General Introduction

The reduction i.e. removal of oxygen at 2 prime of ribose takes place at NDP (nucleoside diphosphate) level. The final step in the sequence of reactions leading to dNDP formation is catalyzed by ribonucleotide reductase (RNR). NADPH (formed in OPP pathway) supplies two hydrogen atoms which finally react with the oxygen at 2 prime of ribose giving water. There are still two more important processes to be discussed. (a) How ribose sugars are converted to deoxyribose so that dNTP's can be formed which can be incorporated into DNA? (b) Secondly, how thymine nucleotide i.e. dTTP is formed?

Reactions

In this process NDP's are converted to dNDP's. These dNDP's can be further phosphorylated to form dNTP's by NDP kinase. Intermediate reactions involve reversible oxidation and reduction of -SH and S-S- groups in thioredoxin or glutathione and glutaredoxin and finally ribonucleotide reductase. In plants as thioredoxin plays a role in activating many enzymes in Calvin cycle it is possible that it may have a greater role.

In *E.coli* both the above routes have been demonstrated. Route 1- RNR in reduced form has 2 SH groups which are used to react with

2 prime oxygen of ribose giving water and dNDP. In this process RNR is restored to its oxidized form and has a disulfide bond (-S-S-). It is reduced by a protein thioredoxin in its reduced form containing 2SH groups and thioredoxin becomes oxidized i.e. a disulfide bond is formed. Oxidized thioredoxin in turn is reduced by NADPH + H^+ in a reaction catalysed by thioredoxin reductase which contains $FADH_2$ as a prosthetic group.

Route-2 -It was observed that mutants lacking thioredoxin can survive. This gave an impression that some other reducing agent(s) must be present to cooperate with reduced RNR and accomplish conversion of NDP's to dNDP's with the regeneration of oxidized RNR. These reducing agents were found to be glutathione (a tripeptide-gamma-glutamyl cysteinyl glycine) which can exist in both reduced and oxidized forms in relation to -SH groups and glutaredoxin a small protein whose -SH groups are interconvertible with -S-S-. Hence this protein also can be said to exist in both reduced and oxidized forms. So in this route RNR disulfide bond is reduced by reduced glutaredoxin and reduced glutaredoxin is oxidized back by reduced glutathione (depicted as 2GSH because -SH group of cysteine from 2 molecules of glutathione interact to form a disulfide bond and when disulfide bond is reduced we get back two molecules of glutathione). Oxidized glutathione is in turn reduced by NADPH +H^+ by the enzyme glutathione reductase. Hence NADPH + H^+ is the original reducing agent -donor of two H - which reduces glutathione which in turn reduces glutaredoxin which in turn reduces RNR which finally interacts with NDP's and convert into dNDP's and water and oxidized RNR is regenerated. A tyrosyl radical is involved in the reaction. (The exact mechanism of RNR action, students may learn from other advanced courses).

Regulation of RNR

RNR is the rate limiting step for DNA synthesis. Its regulation is very important to provide the required amounts of all the four dNTP's to preserve genome integrity. Any deficiency or excess can affect genome integrity or can even be lethal. RNR is a target for anti-cancer and antiviral drugs. Three classes of RNR are found. RNR I present in almost all eukaryotes and many prokaryotes and consists of two different subunits, a larger (alpha or RR1) and a smaller subunit (beta or RR2). RNR II and III are found only in prokaryotes. They are found as monomers or dimers but RNR I can be found as higher oligomers as well-tetramers and hexamers. In *E.coli* the catalytically active form is a hetero tetramer-2 subunits of each type and becomes inactive if dissociated into dimers. So RNR I catalytic activity is controlled by its oligomerisation status. Details differ considerably in different systems. In mammals, Class I RNR can exist as dimers, tetramers or hexamers. As monomers RNR is inactive. It has three sites *viz.* active site, specificity site (together referred to as regulatory sites) and hexamerisation site. Initial binding to the specificity site of any of the following nucleotides *viz.* ATP, dATP, dGTP or dTTP favours the formation of dimers which can be catalytically active. Further oligomerisation into a tetramer form occurs because of the binding of dATP or ATP to the active site. The major catalytically active form is a hexamer. Its formation is facilitated from the tetramers formed as above by the binding of ATP to a specific site called hexamerisation site.

Both in *E.coli* and mammalian systems RNR (in both systems RNR I is present) is also subjected to allosteric regulation. Depending on the allosteric effector selective reduction of NDP's into dNDP's is favoured. Reduction of all NDP's is inhibited by dATP. Reduction of pyrimidine nucleotides, CDP and UDP reduction is favoured by the binding of ATP but their reduction is inhibited by dTTP and dGTP (dGTP however does not inhibit in *E.coli*). Reduction of pyrimidine nucleotides CDP and UDP reduction is also inhibited by dGTP. Reduction of GDP is favoured by dTTP and dGTP favours reduction of ADP. RNR III act on NTP's to convert them into deoxy form. Further it is favoured under anaerobic conditions and is sensitive to oxygen, whereas oxygen favours the activation of class I RNR.

Formation of dTTP

Till now we have discussed the formation of all nucleotides except dTTP. dTTP synthesis is one of the important pathways essential for DNA synthesis.

Both CDP and UDP can be converted to their respective deoxynucleotides, dCDP and dUDP by RNR. These dNDP's can then be converted to their respective triphosphates dCTP and dUTP by NDP kinase. dUTP can also be formed as a result of deamination of dCTP by a deaminase. dUTP formed as above can be hydrolysed to dUMP by dUTPase (PPi is the other product). dUMP so formed is reductively methylated by the enzyme thymidylate synthase (TS) to give dTMP. This reaction may be represented as:

dUMP + N5-N10 methylene tetrahydrofolate $\rightarrow$ dTMP + 7,8 Dihydrofolate (DHF).

dUMP so formed can be successively converted to dTDP and dTTP by respective nucleotide kinases. For the continuous synthesis of dTMP, N5-N10 methylene tetrahydrofolate has to be regenerated. This regeneration takes place by a series of two consecutive reactions:

(a) Dihydrofolate reductase (DHFR) catalyzes the reaction- 7,8 DHF + NADPH+ H^+ tetrahydro folate (THF) + $NADP^+$.

(b) Serine hydroxy methyl transferase (SHMT) reaction regenerates N5-N10 methylene- tetrahydrofolate from THF as follows:

THF + Glycine $\rightarrow$ Serine + N5-N10 methylene tetrahydrofolate.

N5-N10 methylene tetrahydrofolate so formed can once again participate in the thymidylate synthase reaction and give rise to dTMP for further formation of dTTP. (SHMT, has been discussed under chapter 14).

Thymidylate synthase is a very important enzyme not only for the synthesis of DNA but it is also a target for drugs to treat many types of cancer. Further DHFR is also a target for cancer chemotherapy by the drugs like methotrexate, aminopterin (both inhibit mammalian DHFR) and trimethoprim (inhibits bacterial DHFR). Resistance to these drugs can occur because of DHFR gene amplification and thereby increased amounts of DHFR.

Part - VI

Regulation of Few Important Metabolic Pathways in Plants

23

Calvin Cycle-C3, C4 and CAM Pathways

General Introduction about photosynthesis

Photosynthesis -the conversion of light energy into chemical energy can be carried out by plants, algae and prokaryotes. The process ultimately supports the heterotrophic organisms. It supplies oxygen for aerobic organisms to survive and maintain atmospheric oxygen concentration and supplies energy and reducing power for the synthesis of food material that can be consumed by animals and humans and microorganisms. Some prokaryotes can also carry out anoxygenic photosynthesis i.e. where oxygen is not evolved. Our primary concern is with oxygenic photosynthesis where water is used as the ultimate source of reductant. It is the most important process to manipulate for increasing crop yields to feed the large and still growing population.

Photosynthetic process is subdivided traditionally into two broad types of metabolic reactions—light dependent reactions and dark reactions also called light independent reactions. However, modern knowledge has revealed that light reactions are also essential for the dark reactions to operate and hence, strictly speaking, they are not dark reactions but dependent on light. Energy in the form of ATP and reducing power in the form of NADPH + H^+ (*via* reduced ferredoxin) is used for further reactions to take place which is called as Calvin cycle. This cycle is also called as Calvin-Benson cycle or

Calvin-Benson-Bassham cycle or reductive pentose phosphate pathway -hereafter just referred to as Calvin cycle. Further light induced increase in the pH and divalent Magnesium ions and formation of reduced thioredoxin *via* ferredoxin is essential for activating the enzymes of Calvin cycle. Both light reactions and Calvin cycle take place in the chloroplasts -light reactions in the membrane systems (thylakoid/grana) and Calvin cycle and other metabolic reactions in the chloroplast stroma. In this process chloroplasts actively interact with the cytosolic compartment for the synthesis of carbohydrates apart from synthesizing carbohydrate to be stored in chloroplast. Many enzymes of Calvin cycle also operate in OPP pathway and gluconeogenesis. In fact, many reactions of Calvin cycle also take place in animal cells.

Chloroplasts number varies widely per cell from 1 in algae up to 100 in plants like Arabidopsis and wheat. In general, in leaves they may vary from 50 to 100. Lower numbers also are reported. In turn in each chloroplast may contain up to 100 copies of the genome. Hence there can be a large number of copies of different genes getting expressed cumulatively contributing to higher amounts of RNA and/or proteins.

The first and the most important reaction in the Calvin cycle is the carboxylation of Ribulose-1,5,-bisphosphate (RuBP) by the enzyme ribulose-1,5,-bisphosphate carboxylase/oxygenase. The enzyme is abbreviated differently by scientists but here the abbreviation "Rubisco" is used. This enzyme deserves special attention for many reasons. (a) It is the most abundant enzyme/protein on earth. It constitutes nearly 50 per cent of soluble protein in leaves amounting to 20 to 30 per cent of leaf nitrogen in case of C3 plants. Its abundance is lower in case of C4 plants; 5 to 30 per cent of soluble leaf protein. (b) It is a very large enzyme consisting of 8 large sub-units -each of approximately 55,000 Da and 8 small sub-units -each of approximately 13,000 Da. Hence in all it has 16 sub-units designated as L8S8. The large sub-unit is coded by the chloroplast genome while the small sub-unit is coded by the nuclear genome. Small sub-unit is synthesised in the cytosol and transported into the chloroplast matrix. In some prokaryotes it is just a dimer of two large sub-units. (c) It is a very inefficient enzyme with a low catalytic turnover rate of 3 to 10 molecules per second. How low it is, can be appreciated by comparing the value of 5 lakh molecules per second for the enzyme carbonic anhydrase. (d) It is the only enzyme fixing net carbon dioxide with the regeneration of the CO_2 acceptor molecule RuBP. Hence it is an auto catalytic cycle. (e) It not only fixes CO_2 but also act as a storage protein. During grain/seed development the leaves undergo senescence when the Rubisco is degraded and the amino acids supplied to the developing grain/seed as a source of carbon and nitrogen. (f) In addition to fixing CO_2 it also fixes oxygen- a process called C2 cycle or photorespiration which is considered as an inevitable reaction. Photorespiration is considered as a wasteful process energetically and reduce net photosynthesis. Where only CO_2 reacts the sequence of reactions are referred to as C3 cycle. (g) It is the most important regulatory step. Rubisco catalyses the following reactions.

(i) Carboxylase reaction-RuBP + CO_2 + H_2O $\rightarrow$ 3-phoshphogly cerate + 3-phoshphoglycerate i.e. two molecules of 3-phoshphoglycerate (3PGA)

(ii) Oxygenase reaction- RuBP + O_2 $\rightarrow$ 3PGA + 2-phosphogly colate (2PG)

Since 3PGA is a 3-carbon compound its further metabolisation is called a C3 cycle, while 2PG is a C2 compound and hence C2 cycle. 3PGA is a triose phosphate and can be isomerised into another triose phosphate, dihydroxyacetone-3- phosphate (DHAP). C2 cycle ultimately leads to loss of energy and loss of one carbon in the form of CO_2. This cycle is called photorespiration. As triose phosphate contains 3 carbons, 3 rounds of net CO_2 fixation is required to generate it, while glucose which has 6 carbons needs 6 rounds of net CO_2 fixation. Six round give 12 molecules of 3PGA i.e. 36 carbon atoms. Six carbons are equivalent to glucose and so thirty carbons are remaining. These 30 carbons are rearranged to form RuBP and 6 RuBP initially involved in the fixation of 6 CO_2 are regenerated. To fix one molecule of CO_2, 3ATP and 2 NADPH + H^+ are required.

Regulation of Photosynthesis/Calvin cycle (C3) C4 and CAM and photorespiration pathways

Though we have C3, C4 and CAM pathways. Detailed discussion here is confined to regulation of C3 cycle with a brief overview of regulation of the other pathways including photorespiration. In other pathways -though initial CO_2 fixation process is different, ultimately CO_2 fixation takes place by C3 pathway i.e. Calvin cycle.

(a) Both amounts and activity of Rubisco are regulated. As mentioned above the number of chloroplasts per cell, number of genomes per chloroplast and the amount of Rubisco can vary. Further as mentioned above both chloroplast and nuclear genomes are involved in coding the large and small sub-unit respectively. Hence there has to be a perfect coordinated regulation of these two different sub-units and they need to be properly assembled in the chloroplast matrix. Small sub-unit is apparently reaches the matrix from cytosol after the cleavage of signal sequences. Catalytic and substrate binding sites are located on the large sub-unit. Role of small sub-unit is not so clear but is implicated in playing some regulatory role. In the assembly of sixteen sub-units some folding proteins (chaperones) may help.

Methyl jasmonate (see signal transduction) can cause alternative processing of the large subunit of Rubisco. Alternative processing leads to the formation of transcripts with a longer (94 bases) 5' untranslated region (UTR) and a shorter transcript which is considered as normal form having 59 bases UTR. The longer form inhibits translation. Methyl jasmonate may also destabilise some transcripts of Rubisco small sub-unit.

In the night when there is no photosynthesis, a metabolite -2-carboxy arabinitol-1-phosphate (CA1P)- is formed from F-1,6-bis phosphate (F1,6BP) which is an intermediate in Calvin cycle. CA1P resembles the transition state metabolite formed in the Rubisco reaction (2'-carboxy-3-keto-D-arabinitol 1,5, -bisphosphate (CABP-beta ketoacid intermediate) which tightly binds to the enzyme and makes it inactive. Further in dark other enzymes undergo protein degradation. In light a molecule of CO_2 bind to the positively charged epsilon amino group of lysine -201 residue to form a negatively charged carbamylate. This in turn binds to divalent magnesium ions to form a catalytically active ternary complex. Another enzyme rubisco activase facilitates the formation of carbamate/carbamylate and also removes the inhibitor CA1P from Rubisco. CA1P so formed is hydrolyzed by a phosphatase to form carboxy arabinitol and Pi. Distinguish the CO_2 bound to lysine with the CO_2 acting as substrate. CO_2 bound to lysine is called activator CO_2 also called non-substrate CO_2 and is different from substrate CO_2 which binds to an enediol intermediate in the Rubisco reaction giving two molecules of 3PGA. Magnesium binds both to active site and the allosteric site i.e. to the carbamate. Two isoenzymes of Rubisco activase are found and hence isoenzymic control may be there. One of the isoforms is sensitive to ADP inhibition.

(b) Two other major light induced changes are increase in stromal pH from 7.0 to 8.0 and increase in divalent magnesium ion concentration from 3 to 6 mM. These two changes along with increased formation of the reducing NADPH + H^+ stimulates several enzymes. Alkaline pH also favours the formation of carbamate. Enzymes which are stimulated by alkaline pH and increased magnesium ions are; Rubisco, F1,6, B Pase and Sedoheptulose 1, 7, Bisphosphatase (SBPase). Purified Phospshoribulokinase from Brassica was inhibited by RuBP, 6-phosphogluconate and AMP, and activated by glucose -1-phosphate, glucose -6-phosphate and inorganic phosphate. This is the enzyme involved in the regeneration of RuBP from R5P:

$$\text{R-5-P} + \text{ATP} \rightarrow \text{RuBP} + \text{ADP.}$$

NADPH+ H^+ formed during the light reactions stimulates the activity of several enzymes by interconversion of thiol-disulfide interconversion. Oxidised thioredoxin (with -S-S-, disulfide bond) is reduced by reduced ferredoxin formed in the light reaction by the action of the enzyme Ferredoxin-thioredoxin reductase. Reduced thioredoxin in turn can reduce disulfide bonds in several enzymes and make them active. Enzymes which can be activated as above are:

F1,6, BPase, SBPase, phosphoribulokinase (PRK), NADP-GA3PDH, Rubisco activase, ATP synthase.

The above enzymes catalyze reactions as follows:

(i) F1, 6, B Pase -F1, 6, BP + $H_2O \rightarrow$ F6P + Pi;

(ii) SBPase-SBP + $H_2O \rightarrow$ S7P + Pi;

(iii) PRK -Ru5P +ATP $\rightarrow$ RuBP + ADP;

(iv) NADP-GA3PDH-1,3, BPG + NADPH + $H^+ \rightarrow$ GA3P + NADP + Pi;

(v) Rubisco activase reaction is described above.

(vi) ATP synthase facilitates the formation of ATP from ADP and Pi during light reactions.

The oxidized thioredoxin formed as a result of reducing the enzymes can in turn be reduced by another molecule of reduced ferredoxin. In the night the enzymes become inactive both by getting reconverted to disulfide forms and by getting degraded by proteolysis.

Thioredoxin is also found in cytosol and mitochondria in addition to chloroplast and hence can regulate several enzymes in different metabolic pathways in these compartments. A few examples are; Phenyl alanine ammonia lyase (involved in phenolic and lignin biosynthesis; G6PDH in the OPP pathway, GA3PDH which is also involved in glycolysis and gluconeogenesis; NADP-malate dehydrogenase in the C4 cycle. Involvement of thioredoxin in the formation of deoxyribose sugars has been discussed under nucleotide regulation. Thioredoxin has also been found to stimulate fatty acid biosynthesis and translation.

(c) Triose phosphate tanslocator (TPT) is an important regulatory step and is involved in exchanging three carbon compounds of the chloroplast with Pi of cytosol. Overexpression of chloroplast TPT in Arabidopsis increases CO_2 assimilation rates in moderate and elevated light conditions. The growth enhancement in Arabidopsis also requires the simultaneous over expression of cytosolic FBPase.

Transgenic potato plants with reduced expression of the TPT achieved by antisense inhibition approach. These plants show severe reduction: (i) in the import of inorganic phosphate from cytosol (ii) of photosynthesis (iii) of plant growth (iv) of sucrose and amino acids accompanied by a parallel increase in the content of starch.

(d) Photorespiration is regulated by compartmentalisation as the reactions take place in chloroplast, peroxisomes and mitochondria. Hence metabolite flow has to be regulated. The oxygenase activity of Rubisco as mentioned above is a wasteful process. On an average 25 per cent of fixed CO_2 is lost in photorespiration. In general ratio of carboxylase activity to oxygenase activity is 3:1. However natural variation occurs in the kinetic parameters i.e. K_m and V_{max}. Among the land plants the ratio of the rate of carboxylase to oxygenase activity as represented by the equation (Vc/Kc)/(Vo/Ko) called specificity factor may vary from 80 to 130 with average being 100. In the above equation $V=V_{max}$ and $K=K_m$ and c=carboxylase and o=oxygenase. Attempts to abolish oxygenase activity have not met with success apparently because the binding sites for O_2 and CO_2 overlap a lot on the enzyme. Some think that there is an adaptive advantage for the photorespiration. At high light intensities and high temperature and moisture deficient conditions, high energy production may damage the photosynthetic apparatus-photooxidative damage. Under such conditions high photorespiration can help dissipating excess energy and prevent damage to photosynthetic apparatus. However, the exact consequences and benefits can be known only when scientists are successful in redesigning the enzyme to possess only carboxylase activity and no oxygenase activity.

In nature plants have evolved mechanisms to avoid photo-respiration. There are two broad mechanisms:

(e) C4 pathway (f) CAM (Crassulacean acid metabolism) pathway.

(e) C4 pathway- They constitute just approximately 5 per cent of terrestrial plants. These plants e.g. Maize, Sugarcane increase the local concentration at the site of Rubisco and increase the rate of carboxylation and avoid photorespiration. When CO_2 concentration is increased carboxylase activity increases and oxygenase activity decreases and vice versa. These leaves have a specific anatomy called Kranz anatomy where two types of cells -mesophyll cells and Bundle sheath cells collaborate in the operation of photosynthesis. In these plants the initial fixation of CO_2 is carried -not by Rubisco- but by another enzyme Phosphoenolpyruvate (PEP) carboxylase. PEP in mesophyll cells combine with CO_2 to give oxaloacetate and Pi. Since the initial product of CO_2 fixation oxaloacetate is a four carbon compound this pathway is called C4 pathway. Oxaloacetate so formed in different plants can be converted to either malate or aspartate and transported to Bundle sheath cells where Rubisco and other Calvin cycle enzymes are present. These metabolites get decarboxylated in bundle sheath cells and increase the local concentration of CO_2 increasing the carboxylation efficiency of Rubisco. Even by any chance if CO_2 is released in bundle sheath cells that is trapped again in mesophyll cells by the enzyme PEP carboxylase and shunted back to bundle sheath cells. Hence no loss of CO_2. PEP regeneration takes place in the mesophyll cells from the metabolites transported from bundle sheath cells. Brief description of the regulation of main enzymes is given here. PEP carboxylase is activated by reduction of disulfide double bonds into thiol (sulfhydryl groups) as mentioned above. Malate and aspartate causes feedback inhibition of the enzyme. They also inhibit the enzyme by reducing the pH. Subsequently these are transported out of the cytosol. Glucose-6-phosphate activates the enzyme and can also counteract the inhibition by malate. NADP-MDH is activated by light, in the same manner as mentioned above for other enzymes in C3 cycle by the ferredoxin/thioredoxin system. Pyruvate inorganic phosphate dikinase (PPDK) can regenerate PEP from the transported pyruvate. It catalyzes the reaction:

$$\text{Pyruvate} + \text{ATP} + \text{Pi} \leftrightarrow \text{AMP} + \text{PEP} + \text{PPi}.$$

PPDK is an important enzyme regenerating PEP for CO_2 fixation and may likely to function as a storage protein too during seed development, as it is abundant constituting up to 10 per cent of the chloroplast soluble protein fraction. PPDK is regulated by the bifunctional enzyme PPDK regulatory protein (PDRP). In the presence of light PPDK is activated by dephosphorylation. Threonine residue at the 456th position undergoes dephosphorylation by the activity of PDRP. The enzyme can be deactivated by phosphorylation of the same threonine residue and by the same protein PDRP. PPDK is also regulated by oligomerisation status. It is active as a homotetramer while it's dimeric and monomeric forms are inactive. Tetramerisation requires divalent Mg^{2+} ions. The activity of Phosphoenolpyruvate carboxykinase (PCK) is regulated by reversible phosphorylation and dephosphorylation. In light it become active by undergoing dephosphorylation.

Since there is so much of compartmentalisation is involved -two different types of cells and three different sub-cellular compartments - cytosol, chloroplast and mitochondria- regulation by compartmentalisation is an important mechanism and accordingly regulation of transport of metabolites between cells and organelles.

Under adverse environmental conditions like drought, high temperatures, nitrogen deficiency or lower levels of CO_2, C4 plants are able to more efficiently fix carbon compared to C3 plants. However, in optimum conditions C3 plants can be more effective.

Scientist by using modern molecular approaches like anti-sense technology have reduced Rubisco and other photosynthesising enzymes to a significant extent and found that there has been no effect on photosynthesis. However, these workers have not observed the final yield of the plants. Further, scientists because of the above reasons, using molecular approaches are trying to convert C3 plants into C4 plants. However, there may be a risk involved as it is speculated that this conversion may accompany a severe reduction in the amounts of Rubisco. Rubisco should be viewed not just as an enzyme of CO_2 fixation but also as a storage protein. During grain/seed development the leaves undergo senescence when the Rubisco is degraded and the amino acids supplied to the developing grain/seed as a source of carbon and nitrogen. Hence any reduction in the amounts of Rubisco may decrease grain yield and/or protein content. Hence suitable molecular approaches must be thought of. Whether the presence of Kranz anatomy is obligatory for C4 pathway to function? This aspect also must be seriously thought over.

Not only Rubisco even other proteins are mobilised from leaves into seeds. Hence reducing Rubisco and other Calvin cycle enzymes by antisense technology to a significant extent though may not effect rate of photosynthesis may affect seed yield/protein content.

(f) CAM pathway—This is so named as the pathway was first discovered in the family of Crassulaceae. However, it is found in other families too. An example of crop plant is pine apple. These plants are found in arid conditions where temperatures are high and there is a shortage of water. Under these conditions these plants close their stomata during the day time but open in the night. So CO_2 fixation by PEP carboxylase is fixed in the night to form oxaloacetate and reduced to malate. Malate so formed is stored in vacuoles. During the day time when stomata are closed, malate stored in vacuole is mobilised i.e. decarboxylated and is fixed by Rubisco. So in CAM plants there is a temporal separation while in C4 plants there is a spatial separation for increasing CO_2 fixation by Rubisco. All CAM plants are not obligatorily function as above. Some can function either as C3 or CAM depending on environmental conditions. So CAM may be induced by adverse environmental conditions. PEP carboxylase is active only during the light period. It can be activated in dark by phosphorylation by a specific kinase.

PEP carboxylase is also feedback inhibited by malate in the cytosol.

Since these plants live under arid conditions, external factors like water availability, temperature and light intensity regulate the metabolism. Regulation of PEP carboxylase is apparently similar to those of C4 plants. There are CAM plants which cannot release CO_2 by decarboxylation of oxaloacetate by PEP carboxykinase or by the oxidative decarboxylation of malate by NADP Malic enzyme for the Rubisco reaction as occurs in different types of C4 plants. So there can be plants which use NAD-malic enzyme. This enzyme in mitochondria catalyses the oxidative decarboxylation of malate into pyruvate with formation of NADH + H^+ and CO_2. This enzyme requires Mg^{2+} or Mn^{2+} and is activated by several metabolites *viz.* acetyl CoA or CoA, sulfate, AMP, fructose-1,6- bisphosphate, and fumarate.

(g) The rate of photosynthesis also depends on the extent of sucrose and starch biosynthesis. By increasing their biosynthesis i.e., the capacity of the sink or utilisation of sucrose the rate of photosynthesis can be increased. This is of practical importance in increasing crop yields.

In plants like sugarcane and sugar beet only sucrose is accumulated and not starch while in seeds and tubers of other crops and even fruits like apple etc. starch is accumulated. In general, a decrease in Calvin cycle enzymes favours the partitioning of carbon into starch rather than sucrose. The triose phosphate formed in Calvin cycle can be transported out of chloroplast into cytosol by a specific triose phosphate translocator/transporter (TPT) in exchange for Pi. This Pi is required for Calvin cycle to continue. Triose phosphate in cytosol can be used for sucrose synthesis which is transported to other parts *via* phloem. Hence TPT is an important target for regulation and manipulation. In addition to TPT we also have Xylulose 5 phosphate translocator which can link OPP pathway with RPP pathway i.e. Calvin cycle. Amyloplasts also synthesise starch from imported carbon, but here the translocator is specific for G6P (GPT).

(Formation and regulation of sucrose and starch biosynthesis is discussed below).

As on July 17, 2016, CO_2 concentration in atmosphere is 403.39 PPM. By 2050, it is speculated that CO_2 concentration would increase to about 550 PPM. It is well known that increase in CO_2 concentration up to the above mentioned level increases photosynthesis. In green houses increasing CO_2 concentration and increasing the yield of horticultural crops is an established practice. However, the paradox is we talk about global warming and adverse effects because of increased CO_2 concentrations in the atmosphere. This because there is no adequate plant cover to absorb the increased CO_2 concentration. In other words, it is told that there is no sufficient sink capacity or demand for the increased CO_2 in the atmosphere. Hence massive afforestation and adequate plant cover throughout the year have to be achieved. Environmental scientists recommend reduction in anthropogenic activities and use of clean technologies that can help in the reduction of CO_2 concentration. (of course global warming is not only because of CO_2 but also because of increase in water vapour, methane, nitrous oxide, and ozone. And some more gases. All these are referred to as greenhouse gases (GHG). However, the two main gases responsible for the greenhouse effect are water vapour (H_2O) and carbon dioxide (CO_2).

24

Sucrose and Starch Biosynthesis

Sucrose biosynthesis and regulation

Sucrose is the major form in which carbon is transported from leaves to other parts which cannot photosynthesize, e.g. roots etc. Hydrolysis of FBP by FBPase is the first committed step in the regulation of sucrose biosynthesis; F1,6, BP + H_2O → F6P + Pi.

F6P can also be formed from G6P by isomerisation and G6P in turn can be formed from G1P by phosphoglucomutase. F6P formed as above interacts with the sugar nucleotide UDPG and forms Sucrose - 6-Phosphate and UDP. The reaction is catalysed by Sucrose phosphate synthase. S6P is hydrolyzed to free sucrose and Pi by sucrose 6-phosphate phosphatase. (students should know that in any organism there are phosphatases which do not have absolute specificity for a substrate but act on several phosphate esters and hence have a broad specificity). UDPG can be regenerated for further sucrose synthesis by UDPG pyrophosphorylase :

UTP + G1P → UDPG + PPi.

UTP can be regenerated from the above formed UDP by NDP kinase: UDP + ATP → UTP + ADP. Pi and PPi released in the above reactions can be transported into chloroplast with the help of transporters.

The enzyme of the first committed step, FBPase is allosterically inhibited by F2,6, BP.

(Formation and maintenance of levels of F2,6,BP is discussed under glycolysis). PFK-2 is inhibited by the triose phosphates formed in Calvin cycle i.e. GA3P and DHAP and is stimulated by F6P and Pi. When photosynthesis is high levels of F2,6, BP are low and F1,6, BPase is active. So it forms F6P which is diverted to sucrose synthesis. In case photosynthetic rates are low then F2,6, BP levels are high inhibiting FBPase. This situation favours the formation of starch from the carbon precursors rather than sucrose.

SPS is stimulated allosterically by G6P and is inhibited by Pi. Its activity is also determined by phosphorylation status. Phosphorylation by a specific protein kinase makes the enzyme inactive. Protein phosphatase removes the phosphate and reverses the inactivation making it active. These regulatory enzymes are in turn regulated. Protein kinase is inhibited by G6P while protein phosphatase is inhibited by Pi. Hence G6P and Pi contribute to increased sucrose synthesis.

There is also another enzyme sucrose synthase which catalyzes the reversible reaction:

Sucrose + UDP $\leftrightarrow$ UDPG + fructose. However, this enzyme is thought to function in the direction of hydrolysis of sucrose rather than its synthesis. Its role in synthesis of sucrose—under any special conditions- needs to be investigated. Another common enzyme involved in sucrose hydrolysis is invertase which liberates glucose and fructose.

An understanding of the regulation of sucrose biosynthesis is directly applicable to crops like sugarcane and sugar beet which store sucrose.

Starch Biosynthesis and regulation

In animals glycogen and triglycerides are the storage forms. In plants the storage forms are starch, triglycerides (lipid or oil), fructans and glactomannans. Starch is the main storage component in the seeds of cereals, legumes and in tuber crops like potato and fruits like apple, banana etc. and starch provides nearly 50 per cent of total calories consumed by humans. Hence starch biosynthesis and regulation is of immense importance. G1P derived from Calvin cycle reacts with ATP to form ADPG and PPi. This reaction is catalysed by ADPG pyrophosphorylase and is reversible. This enzyme is the major controlling step in starch biosynthesis. It is activated by 3PGA the first product of Calvin cycle and is inhibited by Pi. 3PGA accumulates when Calvin cycle is active, while Pi accumulates when photosynthetic phosphorylation is reduced. In case 3PGA is not diverted towards sucrose synthesis then also levels of 3PGA are high and can therefore activate this enzyme. ADPG transfers its glucose residue to an oligosaccharide primer and elongate it by one residue by the action of starch synthase. Repeated cycles increase the length of the chain of alpha $1\rightarrow4$ linkages. Starch consists of two components amylose and amylopectin. The linear chain of glucose units linked by α $1\rightarrow4$ linkages is amylose while amylopectin has in addition α $1\rightarrow6$ linkages, called branches. The branches are synthesised by a separate branching enzyme, which catalyses the formation of α $1\rightarrow6$ linkages. Amyloplasts in the developing seeds cannot photosynthesize and so have to import carbon. They do so by importing G6P and the amyloplast envelope has specific transporters for G6P. There seems to be considerable isoenzymic control. Starch synthase (SS) is found in four multiple forms—SSI, SSII, SSIII and SSIV. These isoforms are found both in soluble portion and also partly bound to the starch granule (those which are bound to granule are termed as granular bound starch synthase (GBSS). SS makes amylose and in conjunction with branching enzyme forms amylopectin. Different isoforms of SS vary in their mode of action and can ultimately in conjunction with branching enzyme produce amylopectin with different structures. Such varied forms of amylopectin have industrial applications. Further manipulation of the activities of starch synthase and branching enzyme can lead to the formation of starch with varying ratios of amylose and amylopectin and these different starches also have industrial applications. Mutants with high amylose are referred to as Floury and mutants with extremely low amylose are termed as Waxy. Such mutants are available in different crops e.g. Maize.

Starch degradation

Degradation of starch is not so much under the hormonal control as in case of glycogen but during seed germination α-amylase degrading starch is up-regulated i.e. its transcription is increased by the hormone

Gibberellin (GA_3) and repressed by ABA. Enzyme alpha-amylase acts endolytically i.e. internally and liberates short glucan chains and also the disaccharide maltose. These are hydrolyzed by alpha-glucosidases liberating free glucose. While alpha-amylase is synthesised *de novo* during germination, beta-amylase is preexisting in the seed. Beta-amylase can cleave amylose from the non-reducing end of amylose liberating maltose units. Maltose can be hydrolyzed to glucose by maltase or alpha-glucosidases. GA_3 can also increase alpha-glucosidase activity acting at genetic level. The preexisting beta-amylase in the seed may be activated either by the removal of a short peptide from the carboxy terminal end or by reducing the disulfide bonds by thioredoxin. During seed germination starch -alpha-1→4 linkages may also be degraded by a phosphorolytic pathway, giving G1P and a glucose chain short of one glucose unit. G1P can be isomerised to G6P by phosphogluco mutase. Repeated cycles of this reaction can occur resulting in shorter amylose chains. These can be transferred to another amylose chain by glucosyl transferase (also called D-enzyme) and again subjected to phosphorolytic cleavage. Phosphorolytic action stops when there are just four residues on the branch point. Beyond this it cannot act. Debranching enzyme removes branch points with alpha-1→6 linkages. Since Pi is involved the availability of Pi may regulate the breakdown of starch.

(Students should compare and contrast regulation of starch biosynthesis with that of glycogen. Glycogen biosynthesis and break down has more complex regulation and more details are known)

25

NITRATE ASSIMILATION

General Introduction and Pathway

In addition to carbon fixation assimilation of nitrogen by plants is another important process. After the three elements carbon, hydrogen and oxygen, nitrogen is the most abundant. CO_2 fixation from air by photosynthesis and water (H_2O) absorption through roots from soil contribute to carbon, hydrogen and oxygen. Among the macronutrients Nitrogen is the most abundant. Plants procure nitrogen from roots in the form of NO_3 (nitrate) primarily and in some cases to some as ammonia. Biological nitrogen fixing organisms can fix atmospheric nitrogen into ammonia and supply to legume plants when they are associated with roots in a symbiotic association. These plants form nodules on the roots where the biological nitrogen fixing organisms reside and convert atmospheric nitrogen into ammonia. However, legumes can be grown under non-nodulated conditions also in which case they assimilate nitrogen in the form of nitrate. The form of nitrogen transport also varies under these different conditions. When legumes are nodulated ammonia formed is ultimately converted to ureides (allantoin and allantoic acid) and transported. Legumes under non-nodulated conditions and other crops transport nitrogen to shoots either in the form of Nitrate or amides (asparagine and glutamine)/and other amino acids, important of which are aspartic acid and glutamic acid. Apparently the regulation of ammonia assimilation in nodulated legumes is different and complex as it involves synthesis of purines and their subsequent degradation to ureides.

Nitrate assimilated from roots can be partly or wholly transported to shoots and accordingly the nitrate assimilation process takes place both in roots and shoots or exclusively in shoots (leaves). It is advantageous to assimilate in leaves as they can photosynthesise and can supply energy and reducing power and precursors for the formation of amino acids. Whereas roots have to depend on carbon from leaves which it has to metabolize to provide the above requirements.

Compartmentalisation is involved in nitrate assimilation. In the assimilation of nitrate, nitrate reductase (NR) is the first enzyme and is subjected to regulation in many ways. The enzyme NR is located in cytosol and require two electrons to reduce nitrate to nitrite:

$NO_3 + NADPH + H^+ \rightarrow NO_2 + NADP^+ + H_2O$. Different forms of NR which can use both NADH + H^+ and NADPH + H^+ and specific to each of them are found in different organisms. NADPH + H^+ specific NR seems to be unique to fungi. The other forms are present in plants.

The second step involves a six electron reduction of nitrite into ammonia by the enzyme nitrite reductase (NiR).This enzyme utilises reduced ferredoxin as reducing power and is located in the chloroplast/ plastid; NO_2 + 6 Fd (reduced) + 8 $H^+ \rightarrow NH_4^+$ + 6 Fd (oxidized) + $2H_2O$. As nitrite is toxic it is rapidly assimilated into ammonia. Even ammonia is toxic and therefore is immediately assimilated into amino acids. Process of ammonia assimilation takes place in chloroplast or in nodules in case of nodulated legumes.

Ammonia formed as above is assimilated further by the action of two enzyme glutamine synthetase (GS) and glutamine-oxoglutarate amidotransferase (GOGAT or also called glutamate synthase) as follows:

Glutamic acid + NH_4^+ + ATP $\rightarrow$ Glutamine + ADP + Pi;

Glutamine + NADH + H^+/NADPH + H^+/Ferrodoxin (reduced) + alpha -ketoglutarate (also referred to as 2-oxoglutarate) $\rightarrow$ two molecules of Glutamate + oxidized forms of the reducing power used. Of the two one glutamate is recycled for the continued fixation of ammonia and the other is utilised for the synthesis of other metabolites. GOGAT is present both in cytosol and chloroplast. Hence isoenzymic control and control by compartmentalisation is involved. However, ammonia assimilation takes place primarily in the chloroplast.

Alpha-ketoglutarate for the above reaction is provided by the oxidative decarboxylation of isocitrate by isocitrate dehydrogenase (ICDH). This ICDH is cytosol localised and uses $NADP^+$ rather than NAD^+ as in case of mitochondrial ICDH. Isocitrate can be formed from citrate in the cytosol by the enzyme aconitase. The above two reactions also take place in TCA cycle in mitochondria. Hence, alpha-ketoglutarate can arise either from mitochondria or cytosol, but it has to be transported into the chloroplast for glutamate biosynthesis. So transport of metabolites across the membranes could be a regulatory step in ammonium assimilation in addition to alpha ketoglutarate biosynthesis which supplies carbon skeleton for the synthesis of glutamate. Biosynthesis of aspartate and asparagine is dependent on oxaloacetate. Therefore, the supply of carbon skeletons for amino acid biosynthesis requires increased glycolytic, phosphoenolpyruvate carboxylation and TCA cycle reactions. Aspartate and asparagine formation and regulation has been discussed under aspartate family amino acids in chapter 14. Thus, metabolism of the above mentioned compounds tightly links nitrogen assimilation and carbon metabolism.

(See discussion of regulation of GS and GDH. Here regulation primarily as occurs in animal and prokaryotes systems is discussed).

Regulation

Nitrate reductase is the major controlling point and it has an elaborate regulation.

(a) The substrate nitrate induces this enzyme and also induces the synthesis of nitrate transporters. Nitrate also induces PEP carboxylase and down regulates the synthesis of ADPG pyrophosphorylase. As a result of this carbon from more carbon is diverted from starch biosynthesis towards the synthesis of amino acids and organic acids.

(b) Light induces NR by supplying reductant, carbon skeletons for amino acid biosynthesis and sugar metabolites. They can act at NR promoter increasing the transcription and also increase NR activity. NR can be induced by red light alone and this effect can be reversed by far-red light. Hence phytochrome (a widely distributed plant pigment involved in the regulation of many processes. It is sensitive to light in the red and far-red region of light spectrum) is implicated in the induction of NR. In dark also NR can be induced in case abundant sugar is supplied.

(c) High levels of glutamine reduces the abundance of NR transcripts.

(d) NR is inactivated by covalent modification by phosphorylation by NR kinase at serine 521. For this to occur binding of an additional protein to NR-belonging to the family of 14-3-3 proteins is required (a family of proteins which can bind to a large number of proteins and regulate several process. The name 14-3-3 is derived because of its behaviour -elution and migration-on DEAE-cellulose chromatography and starch-gel electrophoresis. These proteins are eluted in the 14th fraction of bovine brain homogenate and were found on positions 3.3 of subsequent electrophoresis). These modifications not only reduce the activity of NR but also increase the rate of its degradation. This process tends to occur during darkness. And light can reverse the effects. Dephosphorylation by protein phosphatase 2A can reverse the above changes and restore the activity. In the dephosphorylated state 14-3-3 protein cannot bind to NR.

(e) Divalent calcium ions increases NR activity.

(f) High levels of hexose phosphate inhibit the activity of NR.

(g) Certain forms of cytokinin at specific concentrations increases the transcription of NR mRNA. Free base adenine or its nucleoside adenosine has no effect but the conjugation of an isopentenyl chain to them causes the increase in transcription. Zeatin and its riboside increases transcription.

(h) NR is inhibited by its own product ammonia.

(i) Different amino acids partially inhibit the induction of NR. These amino acids are: lysine, glutamine, asparagine, glutamic acid, aspartate. All these exert a negative feedback inhibiting the transcription of NR. Even amino acid analogues, azaserine, azetidine -2-carboxylic acid have similar effect. Amino acids may also affect the uptake of nitrate.

(j) As mentioned above the post-translation modification, phosphorylation can inactivate NR. In case serine 521 is mutated to another amino acid, it cannot be inactivated. Similarly, a deletion of the N-terminal domain deregulates NR.

(k) Diurnal variations in NR activity is observed with the maximum activity during the beginning of the day. In the dark it is inactivated as mentioned above. In activation in the dark may also be due to the initial products of ammonia assimilation -Glutamine, Glutamic acid and asparagine.

(l) Constitutive expression of NR gene by linking the gene to a constitutive promoter show abundance of NR mRNA both during the day and night but activity is found only during the day. Hence, light apparently acts on the translation process and/or NR protein degradation. A naturally occurring constitutively expressing NR has been reported in soybean.

(m) Nitrite reductase genes are generally transcribed in coordination with NR expression. However, a few differences have been reported in some plants. The magnitude of the effects of light, sugars and amino acids can be different on NiR expression. In contrast to NR, NiR mRNA levels in the dark were relatively insensitive to supply of sugars. NiR mRNA levels were less affected by the amino acids glutamate and glutamine than NR mRNA levels. The magnitude of increase in NiR mRNA levels was more in the presence of light compared to NR mRNA levels.

26

Regulation of Phenylalanine Ammonia Lyase

Importance

Though the purpose of the book is to devote itself exclusively to primary metabolism-regulation of only one enzyme-Phenylalanine ammonia lyase (PAL), involved in secondary metabolism is included. This is in view of the fact that this enzyme is the first enzyme that diverts the carbon flow from primary metabolism (from Phenylalanine) to the synthesis of an immense number of phenolics (more than 8000 aromatic compounds detected in plants; lignin is the most abundant aromatic polymer) which constitute a major fraction of plant secondary compounds and play diverse functional roles (growth and development, response to biotic and abiotic stresses, anti-oxidants, UV absorbing compounds, pollinators, therapeutic applications etc.). It is also referred to as the committed step in the biosynthesis of phenylpropanoids. This enzyme is of primary significance in plants. However, it has also been reported in some fungi and bacteria but not in animals. It is worth investigating in animals which also possesses many secondary compounds obtained not just from diet but in all likelihood by biosynthesis too!

Reaction and Regulation

(a) PAL catalyzes the reaction; L-Phenylalanine $\rightarrow$ trans-cinnamic acid + NH_3;

PAL from some plants can also act on tyrosine; Tyrosine → p-coumaric acid + NH_3;

(b) The enzyme is regulated at multiple levels *viz.* transcription, translation, post-translational modification, feedback regulation, protein turnover, isoenzymic control, and sub-cellular compartmentalisation.

(i) PAL is feedback inhibited by its product trans-cinnamic acid which presumably acts as a competitive inhibitor. Trans-cinnamic acid can also reduce the transcription PAL gene. In addition to trans-cinnamic acid several other metabolites in the metabolism of phenolics can repress PAL while many others can reverse this repression. The derivatives of phenylalanine *viz.* p-coumarate, O-chlorocinnamate, caffeic acid and related compounds p-hydroxybenzoate and 2-naphthoate also inhibit PAL activity.

(ii) The enzyme activity is also inhibited because of a protein inhibitor. Ubiquitin mediated protein degradation can reduce its amounts.

(iii) PAL is present in multiple molecular forms. Up to 12 different isoenzymes may be found coded by separate genes. These isoforms may be selectively involved in different metabolic pathways of secondary metabolism. PAL can be cytosolic or plastidic or can be present in endoplasmic reticulum.

(iv) Post-translationally modification by phosphorylation at serine/threonine residues by calmodulin dependent protein kinases can reduce the activity of the enzyme.

(v) PAL can exist as an enzyme complex with several other enzymes involved in the biosynthesis of phenolics.

Part - VII

Zymogen Activation, Protein Targeting and Protein Turnover

27

Zymogen Activation

As mentioned above a large number of proteins are initially synthesised as larger inactive precursors and subsequently get cleaved to give a functional polypeptide. So in these cases the proteolytic cleavages can be limited. Only few common examples are discussed here. The inactive precursors may have the prefix 'Pro'- i.e. Pro-protein or Pro-enzyme or have a suffix-'gen'.

Conversion of trypsinogen to trypsin

The inactive trypsinogen has 245 amino acids. It can be cleaved either by an enteropeptidase or by trypsin and remove six amino acids i.e. hexapeptide, from the amino terminal end. These six amino acids are; valine -aspartate- aspartate- aspartate - aspartate -lysine. The numbering of amino acids in a protein starts from the amino terminal end. So amino acid at the amino terminal is counted as one and counting continues along the chain till the carboxyl end. So in the above case we are left with 245 minus six 239 amino acids-numbered from 7 to 245 in the original molecule trypsinogen (However, in an independent polypeptide numbering again starts from amino terminal end. The resulting polypeptide as above is functional and is termed Trypsin. Trypsin catalyzes the cleavage of bonds adjacent to the carboxyl group of basic amino acids-lysine and arginine- in a protein. If proline amino acid is present adjacent to the above amino acids, then trypsin cannot cleave the bonds.

Conversion of Chymotrypsinogen to Chymotrypsin

Chymotrypsinogen also has 245 amino acids and an inactive precursor. Initially it undergoes a cleavage by trypsin adjacent to Arginine 15, giving rise to two fragments -1-15 and 16-245. This is called pi-chymotrypsin and has proteolytic activity. The 16th amino acid is isoleucine. Subsequently chymotrypsin acts on the above two fragments and remove two different dipeptides-serine 14-arginine 15 and threonine 147 -asparagine 148. As a result, now we have three fragments of different lengths-1-13; 16-146 and 149-245. However, these three fragments are not independent but linked together by two disulfide bonds formed between SH group of cysteine residues. One disulfide bond is formed between Cysteine 1 of the first fragment and Cysteine 22 of the second fragment. The second disulfide bond is formed between cysteine 136 of the second fragment and cysteine 201 of the third fragment. This polypeptide is fully functional and is termed alpha-chymotrypsin. However, its site of action in a protein is different from trypsin. Chymotrypsin cleaves bonds adjacent to the carboxyl groups of amides i.e. glutamine and asparagine. Another requirement for cleavage is there should be hydrophobic amino acids e.g. aromatic amino acids (phenylalanine, tyrosine, tryptophan). It can also cleave adjacent to the carboxy group of leucine and methionine but the rate of cleavage is slower as compared to the above. Note that trypsin is involved in the activation of Chymotrypsinogen to Chymotrypsin.

Conversion of pepsinogen to pepsin

This is of importance in human digestion. The inactive precursor is longer by 44 amino acids at the amino terminal end. The matured and functional pepsin has approximately 325 amino acids. In pigs, it is 327 amino acids in the acidic conditions of the stomach arising because of hydrochloric acid secretion pepsinogen unfolds itself and removes on its own 44 amino acids from the amino terminal end. This situation represents autocatalysis which is able to take place in an acidic environment. The resulting polypeptide shorter by 44 amino acids is fully functional and is termed pepsin. Pepsin so formed can not only hydrolyze other proteins but it can also cleave its own inactive precursor pepsinogen to form active pepsin. The site of cleavage in a protein by pepsin is adjacent to amino terminal group of aromatic amino acids and leucine. However, if a proline residue is present to the left of the above amino acids, pepsin cannot hydrolyze the bonds. (**Note**—in case of trypsin and chymotrypsin the cleavage is adjacent to carboxyl group of specific amino acids. With pepsin the cleavage is next to the amino group of specific amino acids).

Conversion of Pre-pro insulin to functional insulin

Note- here in addition the prefix- Pro we also have an additional prefix Pre- This is because 23 amino acids at the N-terminal end act as a signal peptide and is cleaved in the endoplasmic reticulum giving an inactive pro-insulin which is the zymogen form. The zymogen has 86 amino acids. Pro-insulin polypeptide may be subdivided into three regions- A-chain; B-chain and a C-chain also called connecting polypeptide. A-chain is at the carboxy terminal end and consists of 21 amino acids corresponding to 66 to 86th amino acid of the original pro-insulin polypeptide.

B-chain is at the amino terminal end and has 30 amino acids and correspond to 1 to 30th amino acid in the proinsulin molecule. The C-chain or connecting peptide constitute amino acids from 31 to 65 of the proinsulin and is removed by carboxypeptidase. The amino acid sequence at 31 and 32 is -arginine-arginine and the amino acid sequence at 64 and 65 is lysine-arginine. So next to arginine a trypsin or trypsin like enzyme can cleave. Subsequently carboxypeptidase can remove the connecting peptide or C-chain. When once independent A and B chains are formed their numbering of amino acids is independently done starting from amino terminal end. The two chains are not separated but joined together by two inter-chain disulfide bonds between cysteine 7 of A chain and cysteine 7 of B chain and between cysteine 20 of A chain and cysteine 19 of B chain. Hence fully functional insulin hormone has two chains -A and B of 21 and 30 amino acids (total 51 amino acids) held together by two disulfide bonds.

There are so many other important proteins which undergo zymogen activation for example caspases (cysteine-aspartic proteases- they have cysteine as active site and cleave adjacent to aspartate) involved in programmed cell death and prothrombin, fibrinogen etc. involved in blood coagulation. Examples discussed and mentioned constitute only a small number of the total proteins which undergo the post-translational modification of zymogen activation. However, the above examples are typically given to understand the process.

28

PROTEIN TARGETING

Protein targeting

It is also referred to as protein sorting or protein trafficking. Proteins are present in every part of the cell but protein synthesising machinery is present only in cytosol, hence other parts of the cell have to import proteins from cytosol. Though mitochondria and chloroplasts have protein synthesising machinery, very few proteins required by them are synthesised by their own genomes within the organelles. A large majority of proteins required by them are coded by nuclear genes and have to be imported after cytosolic protein synthesis. Right protein targeting is essential for proper functioning of the cell and any errors in targeting can result in diseases in case of humans. Hence the importance of protein targeting. Protein targeting occurs both in prokaryotes and eukaryotes, the magnitude being more in eukaryotes because of more sub-cellular parts or intracellular compartmentalisation.

Information required for protein targeting is present right in the protein itself either at the N-terminal end or C-terminal end or may even reside within the protein. Composition of the transit peptide and its location may vary with the organelle -mitochondria, chloroplast, peroxisome/glyoxysome, vacuole, nucleus and endoplasmic reticulum - to which the protein has to be targeted. The chain of amino acids at the N and C terminal ends are known as signal peptides or signal sequences or transit peptides. Whereas those amino acids residing within the polypeptide responsible for targeting are referred to as signal patches. Though they are situated at a distance, they come together when the protein is folded. The length of the signal peptide is usually shorter varying from 6 to 30 amino acids and sometimes upto 50 amino acids. In rare cases it can be longer. Upto 136 amino acids has been reported. C-terminal signal peptides are mostly concerned with the targeting of proteins into peroxisomes.

Proteins may be targeted to different organelles either Co-translationally i.e. the synthesis of the concerned protein is not yet completed and still has to go further translation or post-translationally i.e. when once the entire protein is synthesised.

Proteins synthesised in cytosol and on rough endoplasmic reticulum (ribosomes attached to endoplasmic reticulum. Endoplasmic reticulum (ER) free from ribosomes is referred to as smooth endoplasmic reticulum) have different destinations. In general proteins synthesised on rough endoplasmic reticulum are targeted to plasma membranes, lysosomes, golgi bodies and outside the cell. They usually follow a Co-translational mode of targeting. Whereas proteins synthesised in the cytosol are mainly targeted to organelles -nucleus, mitochondria, chloroplasts and peroxisomes. In these cases, proteins are targeted post-translationally i.e. after the completion of synthesis of the protein.

Signal sequence at the N-terminal end has hydrophobic and positively charged amino acids. This sequence is recognised by a signal recognition particle (SRP) -an RNA- multiprotein complex. The SRP-ribosome complex then binds to the SRP-receptor present in the membrane of ER and gets attached to ER. Subsequently, the above complex interacts with translocon which is a multiprotein complex which helps the protein to be transported into the lumen of ER. Once translocated into the membrane of ER or Plasma membrane (PM) the signal sequence is cleaved by a signal peptidase. From the lumen the protein is transported in the form of vesicles pinched off from ER. Initially the vesicles are transported to Golgi complex where the vesicles fuse and can undergo the required modifications. After the modifications are complete the proteins are transported to the respective targets. In some cases, the proteins by the help of their N-terminal sequences remain anchored to the membrane. In such cases the N-terminal sequence is not cleaved.

As mentioned above proteins are imported into chloroplast and mitochondria post translationally. After getting into the organelles the N-terminal signal peptide is cleaved by the organelle peptidases to give the matured protein. Since signal peptide is removed to give the matured proteins the proteins that are imported are also referred to as pre-proteins. In these cases, also a translocon is involved. In mitochondria it is referred to as TOM/TIM complex (translocase of outer membrane/ translocase of inner membrane complex). As mentioned above signal sequence at the N-terminal end of organelle targeted proteins also have hydrophobic and positively charged amino acids. However, depending on the target protein and the target the extent of these amino acids and composition of the signal sequence may vary. It may have more hydroxylated amino acids, phenylalanine and leucine and less of alanine and negatively charged amino acids. Some proteins are needed i.e. imported by both mitochondria and chloroplasts. Such proteins have a higher content of hydrophobic amino acids. Since the above organelles have many compartments/sites like outer membrane, intermembrane space, inner membrane and inner matrix, there can exist more than one signal sequence to specifically target the above.

C-terminal signal sequence is involved in the targeting of proteins to peroxisomes in addition to N-terminal signal sequences. These sequences are very short. In general C-terminal signal sequences are different tripeptides (PTS1-peroxisomal targeting signals 1). In terms of single letter amino acid codes they can be; (S/A/C) - (K/R/H)-(L/A) but usually of Serine-lysine -leucine (SKL). N-terminal signal sequences are referred to as PTS 2 and are in general have a consensus nonapeptide sequence i.e. contains 9 amino acids in the overall sequence context of—(R/K)—(L/V/I)—XXXXX—(H/Q)—(L/A/F). X refers to any amino acid in the sequence while other single letters indicate different amino acids as per the single letter amino acid code.

mRNA targeting

In addition, proteins being directly targeted, they also can be targeted by export of mRNA to different targets where they can be translated into a polypeptide also referred to as mRNA based protein targeting. In this case a Cis-acting sequence element acts as the signal sequence for mRNA targeting/localisation. This element is also called localisation element or zip code. This phenomenon has been observed in a wide variety of organisms both pro and eukaryotes. The cis -acting sequence can widely vary from a few nucleotides to one kb. These sequences are located at the 3' UTR (3 prime untranslated regions). Many proteins bind to these regions and the mRNA is transported as mRNA-protein complexes. The binding of these proteins prevents their translation in the cytosol. mRNA targeting can occur to mitochondria, chloroplasts and peroxisomes.

29

Protein Degradation/Turnover

General Introduction and Importance of the Process

In an organism thousands of proteins are continuously being synthesised and degraded at different rates. It is a normal process occurring in every cell. This process is highly essential for the normal growth, maintenance, differentiation and development of an organism. Proteins required at one stage of growth and development may not be required at other stages or not required in the same amounts. Maybe some are not required at all and some are required in varying amounts. Even when a cell is dividing proteins required at one stage of cell cycle may not be required at another stage of cell cycle and hence need to be degraded. Not only unwanted proteins and unwanted amounts of proteins, even defective/damaged proteins and misfolded proteins have to be degraded. Degradation of protein transcription factors can influence the transcription of so many genes. Proteases play an important role in antigen processing. Proteins are degraded during digestion in animals and utilisation as a carbon, nitrogen and energy source by microorganisms. In plants during seed germination seed storage proteins are degraded in to free amino acids and supplied to the growing parts. Similarly, during seed development proteins from leaves and other parts are degraded into free amino acids and supplied to the developing seed for protein synthesis. Proteases and degradation of proteins can be implicated in so many functions in so many organisms. Genomes of different organisms can encode hundreds of proteases. And function of many of them is yet to be investigated.

Many enzymes, non-enzymatic proteins, peptide hormones, and other physiologically active proteins like neurotransmitters etc. are initially synthesised as larger inactive precursors (referred to as zymogens). These are subsequently converted to the active form by the selective proteolysis of specific peptide bonds. In addition, we also have smaller bioactive peptides. Hence the substrate for proteases can be of varying length starting from a dipeptide to an oligopeptide, polypeptide and to a very large protein. Proteolytic cleavages hence represent an important post-translational control regulating the enzyme amounts in addition to other functions mentioned in the discussion. Proteins involved in metabolic regulation are short lived and thus undergo rapid turnover.

In animal systems after digestion it is very common for di and tripeptides to be absorbed in the small intestine and broken down by peptidases.

The complete degradation of a protein into its constituent amino acids may require the action of several proteases acting successively on different lengths of peptides obtained at different stages of degradation. In special cases like zymogen activation there occurs only a limited and selective cleavage. Abnormal proteolytic activity is associated with many human diseases.

In view of the above, process of protein degradation by proteases is very important. While we know how different proteases are involved in the regulation of several proteins their own regulation is less studied and thus is an important research topic. The rate of degradation of a protein not only depends on the activity and amounts and specificity of the proteases but also on the characteristics of the protein substrate.

Proteases are classified on the basis of several criteria

(a) Based on whether they act at the ends of the polypeptide or internally they may be classified as Exo-proteases and Endo-proteases. Further based on which end of the protein they act, exoproteases may be grouped as aminopeptidases or carboxypeptidases. Endo-proteases may cleave a protein adjacent to specific amino acid residues or can be non-specific.

(b) Based on their optimum pH they may be classified as acid proteases, alkaline proteases and neutral proteases.

(c) Based on the amino acid participating in the catalytic activity they may be classified as—serine proteases, cysteine proteases, threonine proteases, aspartate proteases, and glutamate proteases. Further we have two more types metalloproteases which use a metal ion e.g. Zinc and asparagine peptide lyases which use asparagine. Their mechanism is different from other proteases which simply use water and hydrolyze.

(d) Based on the subcellular distribution they may referred to as soluble or cytosolic proteases, vacuolar proteases, lysosomal proteases etc.

Features of a protein which determines its rate of degradation

Features discussed are in general and there can be exceptions. Possessing these features enhances their possibility of getting degraded more rapidly. (a) larger size of a protein, (b) acidic proteins, (c) abnormal proteins, (d) high amide (glutamine, asparagine) content of proteins (e) more disulfide bonds, (f) lesser glycosylation (i.e. compared to a more glycosylated protein a lesser glycosylated protein is more easily degraded. Not all proteins in the cell need to be glycosylated) and (g) higher content of hydrophobic amino acids. All the above mentioned factors 'a' to 'g' increase the rate of protein degradation.

Specific amino acids in a protein have been correlated with the stability of the proteins i.e. half-lives of proteins. Half-life of a protein i.e. time taken to degrade half of the original amount of the protein can be as short as nearly two minutes. In mammals ornithine decarboxylase has the shortest half-life of nearly eleven minutes. An extremely contrasting example is hemoglobin which remains in the erythrocyte for a long time.

(a) Proteins rich in PEST sequences (i.e. P= proline, E= glutamic acid, S= serine, T= threonine) are degraded faster, (b) amino acids present at the N-terminal (amino terminal) end of the protein is another important factor determining the stability of a protein. This is also referred to as N-end rule; (a) Methionine, glycine, alanine, serine, threonine and valine if present at the N-terminal end contribute to greater stability of proteins whose half-lives can be greater than 20 hrs; (b) Isoleucine, glutamine, tyrosine, glutamic acid, proline, leucine, phenyl alanine, aspartic acid, lysine and arginine are considered to destabilise the protein and their half-lives vary from nearly two minutes to thirty minutes; (c) Leucine, phenyl alanine, aspartic acid, lysine and arginine are considered to be most destabilising with these proteins having a half lives of two to three minutes. Proteins involved in metabolic regulation are short lived and thus undergo rapid turnover and hence may have some of the above features.

Ubiquitin mediated protein degradation

(a) General Introduction

Ubiquitin is a small (8.5 kDa) regulatory protein of 76 amino acids of molecular weight 8.5 kDa is present in all eukaryotes and mediates degradation of numerous proteins and regulates numerous important cellular processes in addition to what are mentioned above like translation, histone function and DNA repair etc. Hence it is one of the most important post-translation modifications. Ubiquitin mediated degradation is an ATP dependent process. It is a highly conserved protein among eukaryotes. Humans and yeast ubiquitin differs by just three amino acids. In a series of three consecutive steps ubiquitin gets linked to the target protein that is destined for degradation. The three steps are catalysed by ubiquitin activating enzyme (E1), ubiquitin conjugating enzyme (E2) and ubiquitin protein ligase (E3). All the three are present in multiple molecular forms.

(a) E1 reaction is -ubiquitin + target protein + ATP → ubiquitin thioester of E1 + AMP + PPi. The C-terminal glycine carboxy group of ubiquitin forms a thioester with -SH group of a cysteine residue in E1.

(b) E2-In the second step E2 transfers the thioester from E1 to form a thioester of E2 (ATP is not involved here) and E1 is regenerated.

(c) E3-It is in this step ubiquitin is conjugated with the target protein. The ubiquitin carboxy group formerly present as a E2-thioester is now linked to the epsilon-amino group of a lysine residue in the target protein to give a ubiquitin-protein conjugate and E2 is regenerated. Since it is not alpha-amino group but epsilon-amino group that is involved the bond formed it is referred to as an isopeptide bond.

The above cycle may be repeated to link ubiquitin to more lysines found in the target protein called multiple ubiquitination. This alone is not sufficient for the target protein to be degraded. Ubiquitin can in turn be attached to another ubiquitin molecule. At least four ubiquitin molecules must be attached together in series for the protein to be efficiently degraded. Up to 50 molecules of ubiquitin can be attached to each other to form a chain of ubiquitin residues referred to as polyubiquitin. The process of adding ubiquitin may be called as ubiquitination or ubiquitylation and forming polyubiquitin as polyubiquitination or polyubiquitinylation. The amino acids cysteine, lysine and methionine are given the single letter codes-C, K and M respectively. Polyubiquitin formation is catalysed by a different ligase. In the formation of polyubiquitin, one of the seven lysine residues in the ubiquitin or the N-terminal methionine of ubiquitin is involved in bonding with another molecule of ubiquitin. C-terminus of an additional ubiquitin, molecule can be linked to the epsilon-amino group one of the seven lysine residues found in ubiquitin or to the N-terminal methionine on the preexisting ubiquitin molecule. This process can repeat many times. As mentioned above upto 50 ubiquitin molecules can be added and minimum four are required for efficient degradation. This is how a chain of ubiquitin molecules are formed. Lysine residues in the ubiquitin involved in bonding are usually K48 and K29. Others include K6, K11 and K63. The other two lysines are apparently of lesser involvement. As methionine involved is at N-terminal it is designated as M1. Regulation of different processes involve the polyubiquitination of different lysine residues. K48 and K29 modifications undergo degradation in the proteasome (alternative spelling is proteosome) complex. Monoubiquitination also is an important process which does not involve degradation of the target protein but otherwise regulate different processes. Monoubiquitination may be involved in activating or inhibiting the function of different proteins, protein targeting of different proteins to different organelles within a cell and protein complex formation. All these are non-degradative ubiquitin modifications.

(b) Degradation in the Proteosome Complex

Ubiquitin linked proteins undergoing degradation is a very common process. Polyubiquitin linked proteins formed as above are degraded in a large 26S -approximately 2000 kD - proteasome complex. It is also referred to as UCDEN (ubiquitin conjugating degrading enzyme). This degradation is an ATP dependent process. The 26 proteasome complex consists of a 20S proteasome (catalytic subunit) and a 19S Cap (regulatory subunit. It is also called PA 700 (PA=proteasome activator and 700 indicating the molecular weight) Six ATPase's are associated with 19S regulatory subunit. The functions of 19S include; (a) exclusive binding to polyubiquitinated proteins as only such proteins are degraded; (b) 20S proteasome can act only on unfolded proteins. 19S helps in the unfolding of proteins and presenting it to the catalytic site of 20S proteasome; (c) Though so many proteins are degraded, ubiquitin though it is a protein, it is not degraded by the proteasome complex. 19S possesses an isopeptidase activity which can cleave ubiquitin molecules from the protein. In this way ubiquitin molecules can be recycled for further use. (In addition to this manner deubiquitination can also occur by the help of deubiquitination proteins (DUB) present in the cell. Hence ubiquitination is a reversible process. But if ubiquitins are removed too early -before the addition of the required number of ubiquitin molecules- then the protein may not be subjected to degradation by the proteasome complex).

20S complex is present both in nucleus and in cytosol and in all archaebacteria. The entire complex hydrolyses the proteins -not completely—but into short oligopeptides of 7 to 9 amino acid residues. These oligopeptides are hydrolyzed into free amino acids by cytosolic exopeptidases.

(c) In *E.coli* many proteins are degraded by LON protease. (LON abbreviation is derived from the fact that when this protease is absent the proteins are not degraded and hence found in intact long form). This protease becomes activated in the presence of defective proteins and proteins which have short half-lives. This protease requires two ATP per peptide bond hydrolyzed. Hence it is an energetically a very expensive process and researchers should investigate if hydrolysis can be carried out without the need for energy. The protease does not completely degrade the proteins but into shorter oligopeptides which are further hydrolyzed by other peptidases. These peptidases do not require any energy.

(d) Proteases regulation

Nearly 2 per cent of all genes code for proteases as inferred from the analysis of several genomes and they are involved in so many processes as mentioned above. Hence the study of their regulation is an important research topic. A proteases can undergo autolysis i.e. one molecule of protein degraded by another molecule of the same. There are numerous inhibitors-protein in nature-for proteases in all forms of life. There are inhibitors selectively for enzymes having different amino acids at the catalytic site. These protease inhibitors by inhibiting proteases can regulate many processes. A family of proteins called serpins (serine protease inhibitors) play an important role in blood coagulation and inflammation and other processes. They can also act on other proteases which may not possess serine as an active site. In plants protease inhibitors act as defense molecules protecting the plant against predators and pathogens. They can be pre-existing in seeds and can be induced in plant tissues by herbivory or wounding. Trypsin inhibitors are found in plants, animals and humans. In humans they have clinical significance. They are also serine proteases and trypsin is involved in the degradation of several proteins in different systems.

Part - VIII

A Brief Introduction to Gene Regulation in Prokaryotes and Eukaryotes and Signal Transduction in Mammals and Plants

30

Gene Regulation in Prokaryotes and Eukaryotes

Few Introductory Comments

Though the book is primarily aimed at metabolic regulation dealing with regulation of enzyme activity in different metabolic pathways, a brief overview of genetic regulation/gene expression is given here. That is, how genes can be activated (gene activation) or repressed (gene silencing). This leads to alteration in the levels of enzyme/protein amounts. (Text books on molecular biology and genetics may be referred to for more details.) In the previous section various modes by which enzyme activity is regulated has been discussed.

Details of both prokaryotes and eukaryotes are discussed together. Hence students be alert while reading.

Eukaryotes have a more complex genetic regulation than prokaryotes.

(a) Alterations in Chromosomal/DNA content of the cells

(a) All cells have the same DNA except germ cells. Germ cells have only half the number of chromosomes compared to the somatic cells. (sex determination mechanisms can be different in different organisms). Human sperms have only half the number of chromosomes compared to somatic cells. Further sperm cells can differ in possessing either an X chromosome or a Y chromosome-called sex chromosomes in addition to the set of autosomes. In adult somatic cells one of the two X chromosomes -maternal or paternal is mostly inactivated. In autosomes also either maternal or paternal copy of the gene may be inactivated; a phenomenon called genomic imprinting. In rare cases situation can be quite contrary. Somatic cells can have much less number of chromosomes as compared to germ cells, e.g. Gall Midge which has only 12 chromosomes in somatic cells but germ cells have 48. So in somatic cells the rest of the DNA/Chromosomes are lost. Though all cells have same DNA, there is a variable expression in different organs and tissues and cells i.e. there is a selective expression of a combination of genes in different organs and tissues. Selective expression of genes occurs during growth and development and can also occur because of external pressures. Hence both mRNA and protein types and abundance varies.

It is estimated that about 10 thousand genes are expressed in all cell types of mammals. However, their levels vary widely. These genes are referred to as housekeeping genes or constitutive genes. The promoters of these genes are referred to as constitutive promoters. Genes which are expressed in a specific cell type are referred to as luxury genes.

(b) There is a complete loss of nucleus in the erythrocytes. Whatever Hemoglobin in these cells is synthesised is from the stable mRNA's produced before the nucleus is lost. In case of plants a whole chromosome or even along with its pair can be lost—a situation referred to as monosomic and nullisomic respectively.

(c) DNA amplification—This process can be selective for certain genes or a whole chromosome or a whole set of chromosomes may be amplified. (i) In Drosophila repeated endoduplication occurs forming a giant polytene chromosome. The degree of polyteny can be as high as 1024 in salivary gland chromosomes, i.e., about 512 in each homolog. In plants we have different ploidy levels like triploid, tetraploid, hexaploid etc. (ii) Genes for ribosomal RNA(rRNA) are present in several copies in many organisms. During frog embryonic development the number can be as high as 2 million copies. (iii) The anti-cancer drug -methotrexate- can induce amplification of the gene Dihydrofolate reductase in mammalian cells. (iv) Many genes in many organisms are found in multiple copies in the nucleus. (v) Another situation exists in organelles, mitochondria and chloroplasts. There are multiple copies of the genome in these organelles. Animal mitochondria may have up to 4 genomes per mitochondria and each cell may have up to 20 plus mitochondria. So there are several copies of the same gene. Plants usually have one master copy of DNA with smaller versions of the master copy. However, in chloroplasts the genome amplification is very high. Plant cells may possess 20 to 40 chloroplasts and each chloroplast in turn may have 20 to 40 copies of the chloroplast genome. As high as 100 chloroplasts and 100 copies also have been reported.

(b) Rearrangement of DNA sequences

Humans can produce up to one million antibodies (protein in nature). An antibody broadly consists of a variable region, constant region and a junction region. The large number of different antibodies are produced by gene/DNA rearrangement in B cells (B lymphocytes). Each B cell has a different DNA rearrangement.

(c) Control of gene expression by controlling transcription and production of mature mRNA

In prokaryotes as there is no nuclear and cytoplasmic compartmentalisation, transcription, translation and degradation of mRNA can occur simultaneously. Further there are no organelles like mitochondria and chloroplasts. (However in eukaryotes, in mitochondria and chloroplasts, transcription and translation takes place in the same compartment for their respective genes) In eukaryotes the above processes are compartmentalised and there are many steps. The production of mRNA can be regulated at transcription level and also at post-transcriptional level-also referred to as RNA processing. Expression of some genes is controlled both at transcriptional and post-transcriptional level while some are controlled only at the post-transcriptional level. Regulation at transcriptional level is most common. At transcriptional level the control can be exerted by the frequency of mRNA initiation, strength of the promoter, methylation status of the promoter (demethylation is required to activate transcription), upstream sequences, transcription factors (specific proteins), rate of transcription and enhancers (regulatory sequences which may exist, upstream, downstream, in the transcribed region or within the intervening sequence and can function in either orientation, activating the transcription). Transcription may also be regulated at the termination sites. Sometimes transcription process may pass beyond the termination sites. Transcription can be repressed by regulatory DNA sequences called silencers which like enhancers can be located at a distance and in either orientation. Specific proteins may bind to them and aid in their functioning. The effect of enhancers and silencers is confined to specific genes and not all genes This specificity is often conferred by another set of DNA regulatory sequences called insulators. In addition, there are enhancer like sequences called locus controlling regions (LCR) which can induce expression of genes. In general, transcriptional regulation occurs at initiation level but in some cases it can also occur at elongation level. If DNA sequences e.g. promoters, enhancers, silencers—are regulating the transcription they are referred to as Cis acting factors/elements or Cis regulating and if proteins bind to DNA- e.g. transcription factors- and regulate transcription they are referred to as Trans-acting factors or Trans-regulating.

The amount of a given mRNA available is also determined by its rate of degradation -also referred to in terms of mRNA stability or half-life of mRNA's or mRNA turnover or mRNA abundance. In general, prokaryotic mRNA's have short half-lives (time required to degrade 50 per cent of mRNA)-2 to 5 minutes. In *E.coli* half-lives in general may vary from 10 seconds to 20 minutes-can be up to 90 minutes. In eukaryotes the variation can be very wide -few minutes to several days.

Some genes can be transcribed and result in the production of multiple mRNA's. Therefore, the transcriptome (complete set of RNA's expressed) can be larger than the total number of the genes in the genome of an organism.

The post-transcriptional controls involve addition of a 7- methylated Guanosine CAP structure at the 5′ end of mRNA and addition of a stretch of Adenosine residues-up to 200- called Poly A tail at the 3 prime end of mRNA. (One rare exception is Histone mRNA's which do not possess poly A tail); mRNA splicing (mRNA can be spliced i.e. cut and joined by removing the introns (non-coding sequences or unexpressed sequences and joining the exons (coding sequences or expressed sequences). Splicing may occur either during or after transcription is completed. (Only matured mRNA's are transported to the cytosol). Such genes are called interrupted genes. The number of genes which are interrupted and which are not, vary with the organism. (Interrupted genes is primarily a characteristic of nuclear genes with exceptions being found in sub-cellular organelles; for example Yeast mitochondrial DNA). The ultimate length of a gene is a function of its number and length of its introns. A gene is much longer because of its introns. Any mutation in introns can result in the alteration of RNA splicing which can result in a polypeptide with a different sequence and also can affect the production of a functional polypeptide. Alteration in RNA splicing because of point mutations (single nucleotide/base changes) in introns is responsible for several human diseases. In these cases, a polypeptide with a different sequence or a shorter polypeptide affecting the function is produced. While mutations in exons affect only the polypeptide sequence. The number of introns and exons can vary a lot among different genes in different organisms. The number of genes also can vary a lot among organisms. For example *E.coli* has 4288 genes while in humans it is estimated to vary from 20 to 25 thousand. The small plant Arabidopsis has 25 thousand genes while some other plants can have much larger genome sizes than human genome. The large genomes are a result of large portions of non-coding sequences. In prokaryotes most of the genome-85-90 per cent - is coded either as rRNA, tRNA or as mRNA(polypeptide). It is speculated that at least 15 thousand genes are required for a free living organism to exist.

Another important post-transcriptional control is alternative splicing i.e. the same mRNA is spliced in different ways and different mRNA's which can code for different proteins are produced. Alternative splicing is very common in human genome. An extreme case of alternative splicing is found in case of D scam gene in Drosophila. This gene can be spliced leading to 38,000 (thirty-eight thousand) different mRNA's! Their function needs to be investigated. This example indicates the potential of an organism to produce a variety of proteins by alternative splicing. Because of the above mentioned phenomenon, proteome of an organism (the sum total of proteins available in the organism) is larger than the number of genes. A special phenomenon called - transsplicing (where two different mRNA's are involved in the splicing reaction) is found in nematodes and trypanosomes. However, this is an exception in higher eukaryotes.

Post-transcriptional changes -RNA processing- also occurs for rRNA and tRNA but are not discussed here as they don't code directly for proteins, though they are involved in protein synthesis process. One important feature to be mentioned is the extensive modification of bases/nucleotides in tRNA.

Further either 5′ end or 3′ prime end or both can be differentially spliced. In addition, mRNA editing (change of one nucleotide into another nucleotide at mRNA level) also can occur. For example, change of cytosine to uracil and adenine to inosine resulting in differentially proteins.

Capping and PolyA tails may prevent mRNA degradation by nucleases and thus ensure higher amounts of mRNA available for translation. PolyA tail is also implicated in translocation of mRNA's from nucleus to cytoplasm. Further control points are modification of bases/nucleotides, transport of mRNA's from nucleus to cytoplasm (specific proteins may be associated with mRNA's in helping the translocation process), rate of degradation of mRNA's in the cytosol, transport of translated mRNA's to specific sub-cellular targets like mitochondria, chloroplast, plasma membrane etc. Since organelle genomes code only for a few proteins required by them, the rest come from the nucleus. That is, nuclear genes are transcribed, transported in to the cytosol and translated and then exported to organelles.

RNA interference (RNAi)

It is also called as RNA silencing and post-transcriptional gene silencing and also as quelling. In this phenomenon short (20-25 nucleotides) regulatory RNA molecules inhibit the expression of genes. This inhibition may be done by any of the following mechanisms:

(a) interacting with DNA and preventing transcription, (b) interacting with mRNA and preventing its translation at both initiation and elongation steps, (c) interacting with mRNA and promoting its degradation. Two types of RNA molecules are involved:

(a) micro RNA's (miRNA's) (b) small interfering RNA's (siRNA's). Both are non-coding RNA's (ncRNA's) which are not translated to any polypeptide. This phenomenon is used to silence or knock out genes. In the nematode (roundworm) *Caenorhabditis elegans* RNAi has been used to inactivate 16,000 genes out of its nearly 19,000 protein coding genes and the behaviour of the worm and function of the genes has been studied.

(a) Micro RNA's (miRNA's)

The primary transcripts for these arise either from the introns of the coding genes or from non-coding genes and a few can arise from pseudogenes as well. The large primary transcripts are transcribed by RNA polymerase II in the nucleus and is termed as Pri-miRNA. This is endonucleolytically cleaved by a RNase III type enzyme called Drosha in the nucleus into Pre-miRNA's about 70 nucleotides forming stem - loop structures (In some organisms like Drosophila, Drosha interacts with an additional protein called Pasha). Pre-miRNA's are transported from nucleus to cytosol. In the cytosol Pre-miRNa's are again endonucleolytically cleaved into about 21-23 nucleotides by an RNAse called Dicer (As mentioned above, here also in Drosophila, Dicer interacts with an additional protein called Loq). The stem -loop structure is still maintained. miRNA binds to RISC (RNA induced silencing complex). In the RISC complex one strand of RNA which is not complementary to mRNA -called passenger strand -is degraded and the other strand which has complementarity to an mRNA called -guide strand—is retained. RISC-associated miRNA then binds to the target mRNA, in general at the 3' UTR(untranslated regions) and sometimes in ORF (open reading frame). Target mRNA is then degraded by the Argonuate also called slicer in the RISC complex. Each miRNA may bind to hundreds of target mRNA's and a single specific mRNA may bind to many miRNA's. In case the pairing is imperfect degradation of mRNA may not take place but its translation would be inhibited.

Nearly ten thousand miRNA's-cumulatively in animals and plants -have been identified using bioinformatic tools. Human genome may code over thousand miRNA's which can target 50 per cent of all mRNA's. In addition, some of them may control other genes. A similar situation is supposed to exist in other mammals.

Roles of miRNA- miRNA is involved in many cellular and physiological processes; for example, cell proliferation, cell differentiation, apoptosis and tissue and organ development and cancer. Each form of cancer is characterised by a specific pattern of miRNA expression. Not only cancer, abnormal regulation of miRNA expression is implicated in a huge number of diseases.

(b) Small interfering RNA's (siRNA's)

They are also like miRNA's are short regulatory RNA's of 20-25 nucleotides with overhanging 2 nucleotides at 3′end. They undergo processing like that of miRNA's. The Argonuate involved is different. Other features that characterise them are: (i) They are produced from long double stranded molecules without any stem-loop structure. However, dicer in some cases can produce siRNA's from small hairpin structures. (ii) They are more predominant in plants and lower animals. However, scientists have been using synthetic double stranded molecules to investigate gene function. These synthetic dsRNA molecules can be introduced into the cells by transfection or can be engineered into *E. coli* and fed to nematodes where they can act. (iii) All the processing steps can take place in cytosol. (iv) siRNA has a perfect match with the target mRNA and therefore it can silence only specific genes/mRNA's.i.e. only one specific mRNA. They degrade the mRNA. (v) In plants a specific polymerase RNA polymerase IV is supposed to be involved in synthesising siRNA's and play a role in imparting resistance to viruses and transposons and in development. (vi) siRNA's have been used to study gene functions in relation to several important physiological and pathological processes like cytokinesis, apoptosis, cell signaling and in relation to diseases. They are being exploited to develop therapies.

(Students are advised to compare and contrast miRNA and siRNA. Both are covered under the term RNA interference (RNAi)).

Antisense RNA's

Antisense RNA (asRNA) is a single stranded RNA complimentary to a messenger RNA transcribed in a cell. In literature they have also been referred to as micRNA (mRNA-interfering complimentary RNA). However, Antisense RNA term is widely used. Antisense RNA's are found in all categories of organisms where they are involved in regulating gene expression. Thus, they are naturally occurring. This phenomenon is thus of common occurrence. Huge number of Antisense RNA's are found in humans and mice. As high as 8 per cent of human genes are supposed to be influenced by antisense transcripts while in mice it is 15 per cent. Within the cell the antisense transcripts can be of two different types; (a) cis-encoded-These are transcripts from the opposite strand of the same genomic locus as the sense RNA. It has a long and perfect overlap with the sense transcript (i.e. the transcript which codes for the protein). Most of the naturally occurring antisense transcripts are of this category. (b) trans- encoded-These are transcripts from a genomic locus different from the sense transcript. Therefore, it is a short and has an imperfect match with the sense transcript.

Borrowing the concept of naturally occurring antisense regulation scientists have attempted to chemically synthesise oligonucleotides of either RNA or DNA which are complementary to a segment of the mRNA's to silence it and investigate its function. The mRNA-antisense RNA oligonucleotide and mRNA-antisense DNA oligodeoxynucleotide forms a duplex thereby inhibiting the translation of the mRNA. Further in the duplex subsequently mRNA is cleaved by an RNAse, leaving the antisense oligonucleotide intact. This oligonucleotide can then bind to another molecule of mRNA and initiate its degradation. This is a very powerful approach. If a 15 base oligonucleotide is chemically synthesised it can target any segment in the human genome which consists of 3 billion base pairs. However, the aim is to target specifically a particular mRNA. Not only using naturally occurring nucleotides but antisense oligonucleotide may be synthesised using chemical analogues of the nucleotide bases.

The above approach has been exploited for several applications and fundamental studies in both plants and animals. Only a few examples have been mentioned here:

(i) antisense oligonucleotides have been used for therapeutic applications. They have the potential to be developed as drugs for different types of cancer, diabetes, Duchenne muscular dystrophy, arthritis etc.

Two drugs are already approved in USA. They are fomivirsen-sold under the trade name Vitravene-as a therapy for cytomegalovirus retinitis and mipomersen-sold under the trade name of Kynamro-as a therapy for a particular form of hypercholesterolemia.

(ii) They have been used to create mutants in certain genes to know their effect on metabolic pathways.

A few examples in plants where antisense approach has been used are:

(iii) Inhibiting ACC oxidase involved in ethylene biosynthesis in mustard. These plants showed increased regeneration potential and has less ethylene production.

(iv) Chalcone synthase (CHS) is a primary enzyme -catalyzes the initial step- involved in flavonoid biosynthetic pathway. It catalyses the following reaction:

$$3x \text{ malonyl CoA} + 4x \text{ coumaryl CoA} \rightarrow \text{naringenin chalcone} + 4x \text{ CoA} + 3x \text{ CO}_2;$$

When antisense CHS was used, petunia flowers have lost their purple colour and produced pale corollas. (However, in another important experiment with petunia to increase the pigmentation was met with unexpected results. It was expected that increasing the copies of CHS would increase the pigmentation. Hence transgenic petunias with more number of copies of the genes were produced. But there was no increased pigmentation. Rather they were either variegated or even colourless. This phenomenon was termed as co-suppression.).

(v) Inhibiting chitinase by antisense increases the plants susceptibility to fungal diseases in Arabidopsis. Hence the role of chitinases in fungal resistance has been established.

Translational and Post-translational control

The primary level of gene expression control is at the level of transcription. However, the amount of functional protein also depends on the number of times a given mRNA is translated, rate of translation, post-translational modifications and rate of protein degradation (also referred to in terms of protein turnover; protein half-lives). The type of proteins synthesised may depend on alternative splicing, use of alternative promoters situated at different sites and termination sites. Translation may also be regulated by the codon usage pattern and availability of corresponding amino acyl tRNA's. An mRNA consisting of codons which have abundant corresponding tRNA's is translated efficiently. Another mode of translation regulation is by the binding of specific repressor proteins to certain specific target regions in the mRNA, thereby preventing translation.

Protein splicing also exists in many organisms. That is, where a large precursor protein is cut and joined together to form a different protein. The corresponding names for exon and intron in case of protein splicing are extein and intein. Inteins are removed and exteins are joined together.

In Eukaryotes three different RNA polymerases carry out transcription: RNA Pol I (transcribing rRNA), RNA Pol II (transcribing protein coding genes and small nuclear RNA's involved in RNA splicing), RNA Pol III (transcribing tRNA's, 5S RNA and small nuclear RNA).

Prokaryotes have only one RNA Polymerase for transcribing all categories of RNA. Further mRNA's can be polycistronic i.e. can code for more than one protein (eukaryotic mRNA's are in general monocistronic i.e. code only for one protein. One important exception is the nematode *C. elegans* where many genes are transcribed as polycistronic mRNA's). Though polycistronic mRNA's code for more than one protein, the levels of different proteins are not the same. Hence the rate at which different genes are translated is different. So a translational regulation occurs which decides the amount of various proteins resulting from the translation of polycistronic mRNA's. As half-lives of different mRNA's can vary a lot (mentioned above), this is another determining factor for the variety and abundance of proteins present in a cell. Prokaryotic mRNA's do not undergo either capping or splicing.

Basic modes of regulation at operon level in prokaryotes

In prokaryotes transcription is basically controlled by two means: negative control and positive control. In negative control a repressor protein prevents a gene from being expressed. In positive control a transcription factor is required for transcription to occur. Depending on the additional factors required these two controls can be further divided into:

(a) negative induction (b) positive induction (c) negative repression and (d) positive repression.

(a) Negative induction

Negative induction e.g. lac operon in the metabolism of lactose (An operon is a functional unit of transcription and genetic regulation. It is a segment of DNA containing adjacently structural genes, an operator gene and a regulatory gene. In other words, it is a group of contiguous and coordinately controlled genes). In this case the repressor protein coded by the regulatory gene forms a complex with and is inactivated by allolactose -called an inducer thereby activating transcription. (though allolactose is the inducer, it should be remembered that Lac operon is concerned with the metabolism of lactose) With some other operons the inducers are different. lac operon is also regulated by another mode called, catabolite repression. Two components required for the catabolite repression are cAMP (cyclic AMP also referred to as second messenger) and CRP (catabolite repressor protein also called CAP-catabolite activator protein). When glucose levels are low, cAMP levels increase. cAMP binding is required for CRP to bind DNA and activate transcription.

(b) Positive induction

Unlike in negative induction, here the regulatory gene product is inactive. It binds with an inducer and becomes a functional activator which can bind to DNA and facilitates transcription. e.g. arabinose operon (However, arabinose operon is regulated in several other ways as well).

(c) Negative repression

In this case the regulatory gene product is inactive as a repressor by itself. However, when it binds to another molecule -Co-repressor - It becomes an active repressor which binds to DNA and represses transcription, e.g. Tryptophan biosynthetic operon. Tryptophan is the Co-repressor. Tryptophan operon is also controlled by another mechanism called attenuation. Mechanism of attenuation is as follows: 5 prime end of mRNA-coding for enzymes involved in biosynthesis of tryptophan- can form two different types of looped structures by pairing differently with two different regions of mRNA. Transcription occurs only when a particular type of loop structure is formed. Which type of loop is formed is decided by the availability of Tryptophan. At the beginning of the mRNA there are two consecutive tryptophan codons. When tryptophan is limiting, naturally tryptophanyl-tRNA also becomes limiting and therefore translation cannot proceed. Under these conditions mRNA forms a specific type of loop structure which facilitates transcription of the operon. In case tryptophan is not limiting then translation occurs for some length till a stop codon is encountered. Under these conditions the mRNA forms a different type of loop structure (a hairpin structure is formed) which prevents transcription of the operon. Therefore, mRNA required for the enzymes of tryptophan biosynthesis is not formed. Attenuation mechanism operates in operons other than tryptophan also.

(d) Positive repression

Here the regulatory gene product by itself is an activator. However, when it binds to another molecule called co-repressor, the activator-co-repressor complex cannot bind to DNA. Hence there is no transcription.

(e) Stringent Response

In bacteria transcription can also be controlled by a phenomenon called stringent control or response. When the bacteria are growing under extremely poor nutritional conditions there is a drastic reduction in the transcription of various RNA's, reduction in the synthesis of nucleotides, carbohydrates and lipids and increase in protein degradation. All these changes are supposed to be controlled by two nucleotides *viz.* Guanosine tetra phosphate (ppGpp) and guanosine penta phosphate (pppGpp). Both are synthesised from GTP. An exception to the above general response is, ppGpp activating transcription of some genes.

Control of Gene expression by DNA modification

DNA methylation status can indicate whether a gene is active or inactive. Active genes are non-methylated or under methylated. The methylation takes place on cytosine resulting in 5-methyl cytosine. This cytosine in usually followed by a guanosine residue. Such pairs are referred to as CpG doublets. If many such doublets occur, they are referred to as CpG islands. Many housekeeping or constitutively expressed genes have such islands where cytosine is non-methylated. Demethylation at promoter site (the site at which transcription is initiated) is require to activate transcription. Methylation of genes can occur as an epigenetic event. Methylation may be induced by external pressures.

Control of Gene expression by Histone modification

Post-translational modification of histones can regulate gene expression. Phosphorylation and acetylation have positive effects while others have negative and/or positive effects. This phenomenon has been shown only in some organisms and for some genes. Hence can be further investigated.

Whether a gene is transcriptionally active or not may be known by subjecting it to DNase I digestion. Transcriptionally active genes are DNase I sensitive.

(a) Acetylation-addition of $COCH_3$ group—to the lysine amino group at N-terminal end of histones—particularly H3 and H4- reduces the positive charge of the histones. Transcriptionally active genes are associated with such acetylated histones. Transcriptionally inactive genes are associated with non-acetylated or under acetylated histones. Acetyl group comes from acetyl-CoA. Enzymes which add acetyl group to histones are called histone acetyl transferases and those that remove are histone deacetylases. In addition to acetylation, histones undergo several other short chain post-translational modifications viz. propionylation, butyrylation, 2-hydroxybutyrylation, succinylation, malonylation, glutarylation, crotonylation and beta-hydroxybutyrylations on the epsilon amino group of lysine residues. Cumulatively they are termed as "acylations". Since they are on lysine also referred to as lysine acylations and they affect several metabolic process as a consequence of changes in chromatin structure and thereby gene expression. The role of the above modifications appears to be different than that of lysine acetylation in changing the gene expression. Unique histone acylations have been found to play a role in spermatogenesis, signal dependent gene activation, tissue injury and metabolic stress. As per Sabari *et al* (2016) 246 histone sites bearing the above modifications have been found. Hence there can be several combinations of post-translational modifications. An important inference from the above modifications is, short chain fatty acids have an important role in histone modifications and gene expression. Enzymes which remove the above mentioned acylations have been found. A particular group is termed as SIRTUINS and several isoforms are available.

(b) Ubiquitin is a small protein of 76 amino acid residues. The Carboxy terminal of ubiquitin can be linked to the free amino group lysines. Ubiquitination on the free amino group/epsilon amino group of internal lysines (i.e. not lysine located at the N-terminus) in histones -particularly H2A reduces the net positive charge on the histone molecule and can lead to repression of genes. Interestingly ubiquitination of the histone H2B stimulates gene expression. Mechanism proposed for this opposite effect of ubiquitination is as follows. Ubiquitination of H2B promotes methylation of the histone H3. This methylation occurs on lysine 4 and lysine 79 (amino acids counted from the N-terminal end). Methylation induces a more open structure of chromatin and thereby activates gene expression.

(c) Like ubiquitin SUMO (small ubiquitin related modifier) is a small protein that can be conjugated to lysine in the histones H3 or H4. The process is called sumoylation. Histone sumoylation results in a more closed chromatin structure and thus is a negative regulator i.e. represses gene expression. Formation of closed chromatin structure is supposed to occur because of the removal of acetyl groups from the histones.

(d) Phosphorylation-Histone phosphorylation at serine 10 of H3 has been shown to stimulate gene expression of different genes. Phosphorylation is supposed to result in acetylation of histones leading to open chromatin structure and thereby stimulate gene expression.

Phosphate group comes from ATP and the enzymes are referred to as histone protein kinases and those that remove phosphate group are histone protein phosphatases.

(e) Methylation- Methylation of histones -H3 and H4- on the free amino group of specific lysine residues, in general, inactivates gene expression. Methylation can also occur at arginine residue. Methyl group comes from SAM (S-adenosyl methionine). Methyl group is added by the enzymes histone methyl transferases and those that remove are histone demethylases.

(f) ADP-ribosylation-One or more ADP ribose groups may be added to either lysine or arginine and facilitate transcription. ADP ribose comes from NAD. Enzyme transferring the group are ADP ribosyl transferases.

All histone modifications are reversible. Hence transcription can be reversibly activated or inactivated depending on the need of the cell. Further histone modifications -in addition to transcriptional control -are also implicated in DNA replication, chromatin assembly, splicing and DNA repair.

Role of Ribozymes

These are RNA molecules possessing different catalytic activities. Their exact role in regulation needs to be investigated. However, their role in different processes has been confirmed. (a) Peptide bond formation in protein synthesis is catalysed by 23S RNA molecule which is a part of the large sub-unit of the ribosome. (b) Another ribozyme called ribonuclease P is concerned with tRNA processing. (c) Many introns possess the ability to splice themselves from their precursor RNA's Introns which are classified as Group I and Group II have this ability. (d) Viroids, virusoids and some plant RNA's have the ability to splice themselves.

31

Signal Transduction in Mammals

General Introduction -Signal Transduction (cell signaling)

Living organisms constantly have to respond to both endogenous (internal) and exogenous (external) signals and accordingly have to respond. The process involves a chain of events termed as signal transduction pathways/cascades. Details of these pathways have been worked out to different extent in different organisms. In the present discussion, the discussion is confined to mammalian systems and plant systems and not microbial systems. Further the discussion is confined to only endogenous (internal signals) and not exogenous (external) signals, with the only exception of a brief discussion on light signaling by photoreceptors in plants.

General Details - Signal Transduction (cell signaling) in mammalian systems

We have studied in different metabolic pathways various types of regulation. All these have been discussed as ways of intracellular communication by signals originating within the cell. However, some of these pathways and several other pathways and phenomenon are controlled by extracellular signals. In these cases, signals come from outside the cell and influence the metabolic pathways inside the cell or phenomenon distant from the site of the signal. So this is a case of intercellular communication. The process is referred to as signal transduction or cell signaling. The external signals are converted to intracellular signals. This is also technically referred as -to transduce or transduced or transduction. The need for transduction arises because these external signals cannot pass through the cell membrane and enter the cell. (for those which can enter for example, steroid hormones, the mode of regulation is different) There are many signal transduction pathways. Instead of using the term pathways the term cascades also is used as it involves a specific sequence of steps till the final physiological response is elicited. They share a common scheme and some steps can be similar. The external signals are also referred to as ligands. In case of hormones the term agonists can be used. Agonists bind to a hormone receptor and elicit a response. Other category of substances called antagonists also bind to receptor but cannot elicit the required response. As they also bind to receptor they prevent the binding of the relevant agonist and prevent the physiological response. The external signal or ligand is also referred to as primary messenger. Signal transduction cascades can act to regulate both the enzyme amounts (long-term or coarse control) and enzyme activity (short-term or fine control). Broadly the signal transduction cascade involves the steps of perceiving the external signal at the outer surface of the cell, amplification of the signal response and inducing a series of reactions within the cell ultimately leading to the needed physiological response. The signal perception and transduction, when once it is no more needed is terminated.

We will be discussing a few examples briefly to get an overview of the signal transduction processes. The external signals or primary messengers we will be discussing are; epinephrine, vasopressin, growth hormone, and epidermal growth factor and insulin. After signal perception i.e. formation of primary messenger/ligand -receptor complex, the information is transduced into the cell initially by a set of other molecules referred to as second messengers which carry out the further steps in the cascade. We will be disusing a few important examples of second messengers *viz.* 3',5' cyclic AMP (hereafter referred to simply as cAMP), 3',5' cyclic GMP (hereafter referred to simply as cGMP), divalent calcium ions, inositol 1,4, 5-triphosphate (IP3) and diacylglycerol (DAG). Other important components of the signal transduction pathways include—7TM receptors, G-proteins, GTP, GDP,

GEF/GRF (exchange factor also called guanine nucleotide exchange factor) GTPase activating proteins (GAP) adenylate cylase, cAMP, cAMP phosphodiesterase, protein kinases, protein phosphatases, beta-arrestin, phospholipase-C and calmodulin.

A brief description of all the above components are given before explaining the process.

(a) Primary messengers/ligands/agonists

(a) Epinephrine (also called adrenaline) and nor-epinephrine (also called nor-adrenaline) are synthesised from the aromatic amino acid tyrosine by the adrenal medulla. Together they are referred to as catecholamines. Hence the receptors which they bind are called adrenergic receptors. There are two classes among this -alpha and beta. Receptor interacting with Epinephrine is classified as beta-adrenergic receptors.

(b) Vasopressin- also known as antidiuretic hormone, found in most mammals and is secreted by posterior pituitary. It is a nona-peptide i.e. has 9 amino acids. However, it is initially synthesised as a much longer polypeptide and undergoes zymogen activation by proteolytic cleavage. In most species the 8th amino acid in the nona-peptide is an arginine residue and therefore referred to as arginine vasopressin. In some other cases lysine replaces arginine and accordingly called Lysine vasopressin. Vasopressin has an internal disulfide bonds between first and sixth cysteine residues.

(c) Growth hormone- a polypeptide of 191 amino acids in humans (Stryer quotes 217 amino acids but many sites on Google quote 191 amino acids). It is also called somatotropin. In humans it is abbreviated as HGH or hGH and is secreted by the anterior pituitary.

(d) Epidermal growth factor (EGF)-It is a polypeptide of 53 amino acids with three intramolecular disulfide bonds. It is formed from a large EGF precursor polypeptide by proteolytic cleavage. Yet another example of zymogen activation.

(e) Insulin-(Discussed under zymogen activation). It is produced by beta-cells of pancreas also called islets of Langerhans.

(b) Nature of signal receptors

Receptors are found in the plasma membrane.

(a) The most common and abundant receptors are the 7TM receptors. Seven refers to the number of times it passes/coils through the membrane bilayer and TM for transmembrane. In other words, these seven helices which pass through the membrane. For the same reason they are called as serpentine receptors. However, the distribution of the receptor is such that it has an N-terminal extracellular domain, C-terminal cytoplasmic domain and a 7 helix transmembrane domain. They are also referred to as G-protein coupled receptors (GPCR) as they activate G-proteins after binding the ligand. They constitute one of the largest protein families with more than 20,000 such receptors and 3 to 4 per cent of the human genome (more than 800 genes are found). They are involved in binding a variety of ligands -For example, catecholamines, photons (Rhodopsin -the protein opsin consists of 348 amino acids bound to chromophore retinal), various odorants and tastants (gustatotry), different neurotransmitters and hormones. Hence they influence a variety of biological functions, like utilisation of metabolic fuels, vision, taste and smell etc. They are implicated in cancer, blood pressure and several other processes.

(b) Growth hormone receptor- It is a large polypeptide of 638 amino acids. Like TM receptors it has three domains. Extracellular domain of 250 amino acids, cytoplasmic or intracellular domain of 350 amino acids and a single helix of the remaining amino acids spanning the membrane. The relevant ligand i.e. growth hormone, can bind to two molecules of receptors forming a dimer of the receptor.

(c) EGF receptor- It is a large polypeptide of 1186 amino acids.

(d) Insulin receptor is a dimer of alpha-beta sub-units and has a molecular weight of about 320 kDa.

(c) Nature of the secondary messengers and other components of the cascade

(a) G-proteins- it is termed G-protein as it bind a Guanyl nucleotide, either GTP or GDP. They are a family of proteins and many forms exist playing different roles. They are heterotrimeric i.e. consist of three different sub-units- alpha-beta-gamma.(However, we also have monomeric G-proteins e.g. Ras, Rho etc.). In the human genome there are about 15 genes coding for alpha sub-unit, 5 for beta and 10 for gamma. Hence, huge number of permutations and combinations are possible and there exist a possibility of a huge number of G-proteins to be present. (However, the most important are only very few: G alpha-s, G alpha-i, G alpha-t, G alpha-q and G-alpha-13). The alpha sub-unit is anchored to the membrane by the help of fatty acids, myristic and/or palmitic at or near the N-terminus. When alpha subunit of trimeric G-protein is released, G-protein remain as a dimer -G-beta-gamma. This dimer is anchored to the membrane by prenylation (isoprenoid groups) either a C15 farnesyl group or a C20 geranyl-geranyl group. The prenyl group is added enzymatically to the cysteine residue (forming a thioether linkage with the sulfur of cysteine) at the C-terminal of gamma sub-unit. A consensus sequence CaaX is present at the C-terminus. "C" is for cysteine, "a" is for an aliphatic amino acid and X is for any amino acid. The gamma sub-unit is farnesylated if X is a glutamine or methionine or a serine but geranyl-geranylated when X is leucine. After prenylation the last three amino acids-aax constituting a tripeptide is leaved by a specific protease exposing the prenylated cysteine. This exposed prenylated cysteine then undergoes a methylation at the carboxy group of cysteine forming a methyl ester. It is the alpha sub-unit that binds the GTP nucleotide. The alpha sub-unit of G-proteins have an intrinsic GTPase activity i.e. they can hydrolyse GTP to GDP + Pi. Hence alpha sub-unit exists in interchangeable GDP and GTP forms. This situation is also expressed as G-proteins existing in bound GTP or GDP forms. However, as stated it is the alpha sub-unit that binds the GTP or GDP. We have GTPase activating proteins abbreviated as GAP's. Further there are guanine nucleotide exchange factor (GEF) also called guanine nucleotide releasing factor which favours release of bound GDP. G-proteins can interact with a large variety of receptor, about 500. Hence they are involved in a large number of pathways.

In addition to heterotrimeric G-proteins we also have monomeric G-proteins or small G-proteins or small GTPase's. These have only the alpha-subunit which can reversibly bind to GTP and GDP and are active when bound to GTP and in this form participate in signal transduction. This alpha subunit also has intrinsic GTPase activity.

Ras is a superfamily of five different family of proteins or GTPases's *viz.* Ras, Rho, Arf, Rab and Ran. Their main function is to regulate cell proliferation, cytoskeletal morphology, vesicular transport, membrane trafficking and nuclear transport (i.e. transport of RNA and protein into and out of nucleus) respectively. Ras and Rho can mediate through serine/threonine protein kinases while Arf by activating ADP ribosyl transferase and phospholipase D. Mutations in Ras are found in many types of cancer.

(b) 3′,5′ cyclic AMP (cAMP)-Formed from ATP by the action of adenylate cyclase, a glycoprotein enzyme. The other product is PPi. Adenylate cyclase is a membrane protein and has 12 membrane spanning helices. Both N-terminal end and C-terminal end are facing towards the cytosolic side.

The active site of the enzyme is constituted from the interaction of two C-terminal intracellular domains; one situated between helix 6 and 7 and one after the last membrane helix. Different regions of the enzyme bind to different forms of G-proteins. Mammals can have up to ten isoforms of adenylate cyclase which can vary in their tissue distribution and in regulatory properties. Some isoforms are activated by phosphorylation while some others are inactivated by phosphorylation. 3′,5′ cyclic GMP is formed from GTP by the action of gaunylate cyclase. In this reaction also the other product is PPi.

(c) cAMP phosphodiesterase (cAMP-PDE's) hydrolyses cAMP to 5AMP. The other product liberated is PPi. PDE's also are found in several isoforms. Both cAMP-PDE's and cGMP-PDE's (hydrolyse cGMP to 5′GMP and PPi) belong to a large family of PDE's. Some forms of cAMP-PDE's can be activated by phosphorylation by different protein kinases.

(d) Protein kinases-Protein kinases are a large family of enzymes which phosphorylate various proteins either activating or inhibiting their activity. Further phosphorylation can also influence the intracellular localisation of the protein and the association of the phosphorylated protein with other proteins. They play a role in different metabolic pathways. Protein kinases phosphorylate either a serine or a threonine or a tyrosine residue at specific positions in a protein.

Accordingly, they are named. (a few can phosphorylate histidine and aspartate residues also). They may also be classified as cAMP dependent, cGMP dependent, calmodulin dependent or cyclin dependent or AMP dependent etc. or independent of any other requirement. The fully sequenced human genome presumably contain 518 putative protein kinases constituting about 2 per cent of the human genome. They are classified into two families: 90 tyrosine (Tyr) kinases (PTKs) and 428 serine/threonine (Ser/Thr) kinases (PSKs)(they can phosphorylate both serine and threonine). Some protein kinases can phosphorylate all the three amino acids, Ser/Thr/Tyr. They are referred to as dual specific protein kinases. Analysis of human proteome reveals that phospho serine, phospho threonine and phosphotyrosine constitute 79.3 per cent, 16.9 per cent and 3.8 per cent respectively. So predominantly it is serine that is phosphorylated. However, there are only 107 putative protein Tyr phosphatases (PTPs) and very few approx. 30 protein Ser/Thr phosphatases (PSPs). In all about 200 protein phosphatases are supposed to be there in the human genome.

A single protein may be phosphorylated by more than one kinase. A protein may be phosphorylated only at one residue or at several residues.

(e) Protein phosphatases- These are enzymes which remove the phosphoryl groups added to the protein by the protein kinases and reverse the effects of phosphorylation on the enzyme/protein—also called phosphoprotein phosphatases. Thus, the effect of protein kinase is nullified and can play an important role in the reciprocal regulation of metabolic pathways. We have several protein phosphatases (about 500). They can remove phosphate group from either one or more amino acid residues. Like protein kinases they are also stimulated by various primary messengers. Majority of the protein phosphatases can be sub-divided into two major groups:

(i) phopshoprotein phosphatases (PPP family). e.g. PP1, PP2A and PP2B.

(ii) metal dependent-Magnesium/Manganese dependent protein phosphatases (PPM family)-e.g. PP2C and pyruvate dehydrogenase phosphatase.

(iii) Aspartate phosphatases-a third group which have aspartate at the catalytic site.

(f) Phospholipase-C -It acts on phospholipids. There are many isoforms. In the present context we refer to beta-isoform of Phospholipase - C. This enzyme cleaves the Phosphotidylinositol 1, 4,5-tris phosphate (PIP2) bound to plasma membrane into inositol 1,4,5-triphosphate and diacylglycerol (DAG). Diacylglycerol remains membrane bound while inositol 1,4,5-triphosphate (IP3) is released from the membrane into the soluble portion.

(g) Calmodulin (CaM) -the term derived from the expression - calcium modulated protein -is a 148 amino acid polypeptide. The protein belongs to a family of proteins called EF-hand family. It was termed as "hand" as the EF has a helix -loop-helix structure which are maintained like the forefinger and thumb of a right hand. Such a structure was originally discovered in parvalbumin. It is these domains that bind calcium ions. Four calcium ions are bound. In this binding seven oxygen atoms coordinate with each calcium, six from the calmodulin protein and one from the bound molecule of water. The regions that bind the four calcium ions are found between: 21-32, 57-68, 94-105 and 130-141 in the sequence of calmodulin. From the above mentioned stretch it is clear that calcium binds to a sequence which is exactly 12 amino acids long. These regions are located between two alpha helices in the EF-hand motifs, the first two regions (21-32 and 57-68) are on one side of the linker region and the other two (94-105 and 130-141) are on the other side. Calmodulin is synthesised in many tissues.

(h) Beta arrestin-a 48 kDa protein involved in signal termination. It terminates the signal by binding to phosphorylated receptor. As a result, the receptor's ability to reduce G-protein activation further increases. There are different isoforms of arrestin protein.

Different Signal Transduction cascades

Broadly three different signal transduction cascades as occurring in humans are discussed here *viz.* adenyl cyclase/cAMP cascade, phosphoinositide cascade and Calcium-calmodulin cascade. In terms of receptor changes due to ligand binding, six types of cascades have been discussed *viz.* three types involving a change in the tertiary structure of the receptor, two types involving a change in the quartenary structure of the receptor and one example where there is apparently no change in the conformation of the receptor but ligand binding and the formation of ligand receptor complex is required for signal transduction to operate.

(a) cAMP cascade

When a hormone like epinephrine (primary messenger) binds to the outer surface of the beta-adrenergic receptor-a 7TM type receptor, it causes an alteration of the conformation of the receptor -a change in the tertiary structure. After the ligand binding the conformation of the cytoplasmic domain of 7TM receptor also is changed. It is this change in conformation that has to be communicated to the cell's interior. This signal binding is converted or transduced into the cells interior. In an inactivated state the heterotrimeric G-protein remains bound to GDP (GDP bound to alpha-sub-unit). The hormone-receptor complex activates G-proteins by facilitating the exchange of GDP bound to alpha sub-unit to get released and exchange with GTP. So now the alpha sub-unit binds tightly to GTP and along with the release of GDP, the G-protein dimer-beta-gamma also gets released, as the GTP bound alpha sub-unit has very less affinity to G-beta-gamma. Now G-alpha-GTP can interact with other proteins and so G-beta-gamma. Even a single hormone-receptor complex can cause the above changes in a huge number of inactive heterotrimeric G-proteins. So we have a huge number of G-alpha-GTP available for further participation in the transduction cascade. In other words, the signal response of a single hormone molecule has been amplified a huge number of times. This is the first signal amplification that occurs in the cascade. As mentioned above there are different G-proteins with several alpha, beta and gamma sub-units. 7TM receptors can mediate response via different G-proteins. In this particular case of epinephrine, the G-alpha is designated as G-alpha-s, "s" indicating "stimulatory".

G-alpha-s sub-unit binds to adenylate cyclase—to the two intracellular cytoplasmic domains mentioned above, changes its conformation and stimulates its activity. The surface of the alpha-sub-unit which was earlier interacting with beta-gamma, now interacts with the enzyme adenylate cyclase. Adenylate cyclase catalyses the formation of a large number of molecules of cAMP from ATP. In other words, the initial signal is once again amplified. So this is the second time signal amplification is occurring in the cascade. cAMP in turn activates protein kinase. This protein kinase is called protein kinase A (PKA). PKA is a tetramer consisting of two regulatory sub-units and two catalytic sub-units -R2C2 structure which is inactive. cAMP binds to regulatory sub-units and frees the two catalytic chains. Now the catalytic chains are active to catalyse the phosphorylation of target proteins using ATP as a phosphate donor. PKA phosphorylates specific serine and threonine residues in different target proteins (mentioned above under glycogen regulation etc.). PKA plays an important role in the regulation of glycogen metabolism. This is an example of regulation of enzyme activities -short-term control-by the hormone. PKA is also involved in regulating enzyme/protein amounts i.e. coarse regulation or long-term control. PKA phosphorylates a transcription factor called cAMP response element binding protein (CREB protein). This protein binds to CRE's and increases the transcription of several genes. Some of the important genes whose transcription is increased are; tyrosine hydroxylase, numerous neurotransmitters *viz.* somatostatin, enkaphalin, corticotropin releasing hormone, nerve growth factors etc. Thus, play an important role in many aspects of growth, development, health and disease of animals and humans.

The above cascade continues to operate unless it is terminated and when the organism does not need the changes any more the cascade needs to be stopped. Complete termination of the signal requires different reactions/steps. As mentioned above G-alpha intrinsic GTPase activity. This GTPase hydrolyses the GTP and converts it into GDP. Now the G-alpha-sub-unit is in GDP bound form. In this form it can reassociate itself with the G-beta-gamma dimer and reform the inactive heterotrimeric G-protein bound to GDP. So, now G-protein is not in a position to activate adenyate cyclase to continue the signal transduction cascade. Prevention of the further activation of G-proteins by the hormone—receptor complex is made possible by, (a) dissociation of the hormone from the receptor, (b) inactivating the hormone receptor complex by phosphorylating it at the C-terminus by a specific kinase-beta-adrenergic receptor kinase (this enzyme does not act when the receptor is not bound to the hormone) and (c) binding of a protein beta arrestin to the phosphorylated receptor.

(b) Phosphoinositide cascade

When a primary messenger like hormone vasopressin binds to 7TM receptors, the hormone receptor complex activates G-proteins as above. Here the G-protein is a different one termed G-alpha-q. It activates beta isoform of phospholipase-C which cleaves the Phosphatidylinositol 1, 4,5-tris phosphate (PIP2) bound to plasma membrane into inositol 1,4,5-triphosphate and diacylglycerol (DAG). Diacylglycerol remains membrane bound while inositol 1,4,5-triphosphate (IP3) is released from the membrane into the soluble portion. In PIP2 the first and second positions of the glycerol group is usually esterified with stearic acid and arachidonic acid respectively. Hence the DAG has these two fatty acids. Beta-isoform of Phospholipase-C has the following modular structure:

N-terminal end- PH domain (Pleckstrin homology)-Four EF hand domains-catalytic domain-C2 domain-G-protein interaction domain-C-terminal end. PH is about 120 amino acid residues at the N-terminal end and it binds to PIP2. The name PH -is derived from the letters taken from-platelet and leukocyte C kinase substrate protein in which this domain was first found. Though the enzyme has four EF hand domains which normally binds to calcium ions, the EF domains here has only a few calcium binding sites. Catalytic domain is referred to as protein kinase -C, which phosphorylates target proteins. Next to catalytic domain is present a C2 domain of 130 amino acids which binds to phospholipid head groups. C2 domain is followed by a G-protein interaction domain, i.e. alpha-sub-unit of Gq. The binding of G-protein activates phospholipase-C catalytic domain. This activation involves the interaction of C2 and PH domains. As mentioned above there are many isoforms of phospholipase-C and some of them lack G-protein interaction domain. Such isoforms cannot activate phospholipase-C.

The two end products of phospholipase-C action -PIP2 (IP3) and DAG act further in the cascade. IP3 causes an increase in the intracellular levels of calcium ions. This IP3 is able to do it by binding to a membrane protein, IP3 gated receptor or IP3 gated channel on the endoplasmic reticulum (ER). ER in muscle cells is referred to as sarcoplasmic reticulum. IP3 receptor is a homotetramer. A minimum of 3 molecules of IP3 need to bound to open the channel and facilitate the release of calcium ions. The increase of calcium ions also activates phospholipase-C and protein kinase-C (PKC). Increase in the intracellular levels of calcium ions can affect different processes, e.g. increases glycogenolysis, increases smooth muscle contraction. (**Note:** Calcium in the present context is a component of Phosphoinositide cascade. Whereas calcium can play an independent role as a second messenger binding to calmodulin and affecting different processes).

When no more needed, the signal transduction cascade is terminated in two ways.

(a) IP3 is a short-lived messenger with a half-life of just few seconds. It is inactivated by successive removal of phosphate groups by phosphatases and convert it into free inositol. Free inositol cannot function in the mobilisation of calcium ions from ER.

(b) IP3 may be further phosphorylated into a tetraphosphate (additional phosphate is at position 3) and then loses its phosphate groups and again form free inositol.

The other end product DAG is also regarded as a second messenger and activates protein kinase C. Protein kinase -C phosphorylates serine and threonine residues in many target proteins and affect many metabolic processes. DAG also has a very short life and is inactivated by its hydrolysis into free fatty acids and free glycerol. And thus DAG signaling is terminated. However, as mentioned above the second position in DAG has arachidonic acid. When it is released on hydrolysis of DAG it acts as a precursor for several physiologically important compounds like prostaglandins which in turn are precursors for other physiologically active compounds, prostacyclins and thromboxanes. Hence the phosphoinositide cascade is not a single cascade but involves additional cascades by various signaling molecules.

(c) Calcium-Calmodulin cascade

Calcium is a widely used second messenger. A variety of signals can induce an intracellular increase in calcium concentration few examples are: corticotropin releasing factor (CRF), Thyrotropin releasing factor (TRF), EGF, platelet derived growth factor (PDGF), acetylcholine etc. which by mediating through calmodulin can influence several processes. The normal level of calcium before induction is about100 nM and after induction it is elevated to about 500 nM. Calcium ions as they can form many interactions can join different segments of a protein and can change its conformation. Six to eight oxygen atoms may be involved in the binding of calcium. Calcium can bind to the negatively charged oxygen atoms of acidic amino acids, aspartate and glutamate and the oxygen atoms of the side chains of the amides glutamine and asparagine. The main chain carbonyl oxygen atoms of glutamine and asparagine also can bind calcium.

Calmodulin and its binding to calcium has been described above. The binding of calcium ions changes the conformation of calmodulin. This conformational change results in the exposure of hydrophobic amino acids to the surface of calmodulin which can interact with several proteins. Calmodulin dependent protein kinases can phosphorylate several proteins. Calcium^{2+}-calmodulin complex can stimulate a wide variety of both enzymatic and non-enzymatic proteins and influence several processes. A few examples are; Smooth muscle contraction, regulation of glycogen catabolism by activation of phosphorylase kinase, interaction with phosphatases, lipid metabolism and role in short-term and long-term memory. Ca^{2+}-calmodulin stimulates the plasma membrane ATPase pump. This stimulation results in lowering the calcium levels to the basal level. The lowering of the calcium levels terminates the signal transduction.

(d) Signal transduction mediated by changes in the quaternary structure of the membrane bound receptors

In the cAMP cascade discussed above, the binding of the ligand changes the tertiary structure of the 7TM receptors. However, in certain cases the binding of the ligand results in a change in the quaternary structure of the receptor which then propagates the signal transduction cascade.

(a) A single molecule of growth hormone (described above) binds to its receptor and promotes dimerisation of the receptor. As a result of dimerisation the extracellular domains of the receptor come together and is accompanied by coming together of intracellular domains of the two receptor molecules. The intracellular domain of each receptor is associated with a protein kinase called Janus kinase 2(JAK 2) which phosphorylates tyrosine residues. JAK2 has the following modular structure:

N- terminal—ERM domain——SH2 domain——Protein kinase like domain (non-functional)—functional protein kinase domain—C-terminal domain. ERM derives its name from the first letters of three proteins; Ezrin, Radixin and Moesin. Ezrin is a membrane protein and radixin is a cytoskeletal protein. Whereas MOESIN is an acronym for membrane organisation extension spike protein. ERM has 300 amino acids and help in anchoring the JAK 2 to the membrane. SH2 is 100 amino acid domain which bind to phosphotryrosine residues through arginine residues. SH stands for Src homology - (Src, pronounced as saarc- a tyrosine kinase) SH domains first observed in Src. The coming together of the two JAK 2's associated with the two receptors brings about a cross phosphorylation i.e. one molecule of JAK2 associated with one receptor molecule phosphorylates the other JAK 2 molecule and vice versa. Without phosphorylation JAK 2 is inactive. Phosphorylated active JAK 2 remain bound to receptor. After phosphorylation JAK 2 gets activated and in turn phosphorylates several target proteins. Mutations in JAK 2 can lead to many metabolic diseases/disorders.

This signaling can play an important role in regulating gene expression. A protein called STAT 5 (STAT is an acronym for signal transducers and activators of transcription) is phosphorylated on a tyrosine residue near the C-terminal end. STAT 5 possess SH2 domains. The phosphorylated tyrosine residues of one STAT 5 molecule binds to another STAT 5 molecule and as a result of this the STAT 5 dimer is highly stable, migrates to nucleus and can bind to specific sites on DNA to regulate gene expression. JAK 2 can also phosphorylate growth hormone receptor as a result of which growth hormone receptor can bind to other proteins and participate in additional signaling cascades.

(b) In the above mentioned example JAK 2 was non-covalently associated with the intracellular domains of the growth hormone receptor. We have situations where tyrosine protein kinases are covalently fused with the receptor. Hence they are referred to as receptor tyrosine kinases (RTK's). Therefore, JAK 2 may be referred to as non-receptor protein kinase. Ligands like EGF, insulin and PDGF bind to these type of receptors. EGF promotes cell growth, proliferation and differentiation. The binding of EGF to the extracellular domain of EGF receptor causes it to dimerize and catalyse cross phosphorylation and activation of protein kinase as mentioned above and continues the signal cascade. EGF promotes cell growth, proliferation and differentiation.

From the above two examples we know that dimerisation is essential for the signal transduction to proceed. However, a different situation prevails in case of insulin receptors.

(c) Insulin receptors are dimers which are disulfide bonded (alpha-beta). That is, they preexist in the form of dimers, but signal transduction can occur only after the binding of the insulin. Hence dimerisation alone may not lead to the propagation of signal. Binding of the signal is necessary. Insulin signal transduction cascade can influence cell growth, cell differentiation and metabolism. An important example of metabolic regulation is increased uptake of glucose and synthesis of glycogen.

After the activation of tyrosine protein kinases, further propagation of signal transduction takes place as follows. An adaptor protein termed Grb-2 binds to phosphotyrosine residues of the tyrosine kinase by its SH domains. Grb-2 then binds to another protein Sos. This binding is facilitated by two SH3 domains. (SH refers to Src homology as mentioned above but SH3 has 60 amino acids) of Sos.These domains bind to proline rich segments of polypeptides. In the present case it is Grb-2. (In other words SH2 and SH3 domains facilitate protein-protein interactions). Sos in turn binds to Ras and Ras in its activated form stimulates serine/threonine protein kinases. These kinases then phosphorylate several target proteins which leads to cell growth. Like in heteromeric G-proteins, Sos by acting as guanylate exchange nucleotide factor (GEF) converts the Ras into an activated form binding with GTP. Similarly, GTPases activating proteins (GAP's) promote GTP hydrolysis and inactivate Ras converting it into a GDP form.

(e) Consequences of defects in signal transduction pathways

This aspect has been studied in detail in humans. As mentioned above the signal transduction process has to be terminated and again switched on according to the needs. If the signal transduction is permanently operating or permanently inactivated, it can lead to diseases. A few examples are; continuous signaling may lead to cholera and cancer and permanently switched off signaling may lead to night blindness and pertussis (whooping cough). Their mechanism in brief is as follows: (a) Specific phosphatases remove phosphate groups and terminate EGF signaling. Hence phosphatases are also termed as tumour suppressors. If phosphatases are not functioning it will lead to abnormal cell growth initiated by EGF. (b) If GTPase activity becomes non-functional then Ras continues to remain active in the form Ras-GTP and thus lead to cell growth. (c) Cholera toxin, produced by the pathogen *Vibrio cholera,* is a dimer. One of the sub-units possesses enzymatic activity which catalyses the transfer of ADP ribose to an arginine residue in a protein. Hence it is a ADP ribosyl transferase. When this attaches ADP ribose to G alpha-s,it continues to remain in a GTP bound form. The active G protein continues to activate PKA which opens a chloride channel ultimately leading to an excessive loss of NaCl and water into the intestines. (d) Whooping cough or pertussis is caused by the organism *Bordtella pertussis*. This toxin also has ADP ribosylating activity. But in this case the G-protein is different -not G alpha-s but G alpha-i.When ADP ribose is attached to this alpha sub-unit it reduces the affinity of G alpha to GTP and therefore it remains in an inactive form G alpha-GDP. This modification inhibits adenyl cyclase, closes calcium channels and opens potassium channels and ultimately lead to the symptoms of whooping cough.

(f) In the above examples the ligands are not capable of passing through the membrane and hence bind to the outer surface of the membrane to a receptor and mediate the signal transduction cascades. However, steroid hormones -produced by adrenal cortex-can pass freely through membranes and enter cytosol. Adrenal cortex produces more than 50 steroids. Depending on their biological activity they are classified into three broad categories; (a) those which affect basic metabolic pathways like carbohydrate, lipid and protein metabolism. These are termed as Glucocorticoids e.g. Cortisol. (b) those involved in the functions of the kidney i.e. the excretion of salt and water referred to as mineralocorticoids e.g. aldosterone and (c) those that influence sexual development and functions like androgens and estrogens.

All the above can pass through the membrane, enter the cytosol and bind to their respective soluble receptor. Subsequently this steroid—receptor complex migrates into the nucleus and function as transcription factors mostly to induce gene expression but in some cases to repress gene expression. Hence steroids can influence the enzyme amounts in a large number of metabolic pathways and influence several biological functions.

Nitric oxide (NO)

(a) General details

It is also called nitrogen oxide, nitrogen monoxide or endothelium-derived relaxing factor (EDRF) is a colourless gas and a free radical - i.e. it has an unpaired electron; hence it is often represented with a dot on the nitrogen atom. In nature it is produced by combustion of fossil fuels and during lightning. However, the effect of atmospheric NO on living systems influencing signaling is not known. In living systems, it is an important signaling molecule and is formed in all types of organisms.

In mammals including humans, NO is an important cellular signaling molecule involved in many physiological and pathological processes.

It made a powerful impact in the branches of neurosciences, physiology and immunology. It is a powerful vasodilator with an extremely short half-life of only 2 to 6 seconds in the blood and is immediately oxidized to nitrite and nitrate. It also plays an important role in fertilisation, differentiation, inflammation cardiovascular signaling and apoptosis.

It is biosynthesised endogenously in mammals from L-arginine, oxygen, and NADPH forming citrulline, NO and NADP, by various nitric oxide synthase (NOS) enzymes (Plants and bacteria form NO by a different set of reactions). NOS oxidizes a guanidine nitrogen of L-arginine releasing nitric oxide in the form of a free radical and citrulline. NOS is found in three isoforms 1, 2 and 3. They are referred neuronal (nNOS), inducible (iNOS) and endothelial (eNOS) respectively, each with different functions. The neuronal (NOS-1) and the endothelial isoforms (NOS-3) are calcium-dependent synthesising only low amounts of NO. Whereas, the inducible isoform (NOS-2) is calcium independent and synthesises higher amounts of NO and this can be cytotoxic. Low levels of NO are useful as it acts as a messenger in various functions like vasodilation, neurotransmission, anti-tumor and anti-pathogenic activities. However, chronic expression of NO can cause pathological conditions, like carcinomas, juvenile diabetes, arthritis etc.

(b) Different mechanisms exist by which NO acts. In general, NO can regulate gene expression in three different ways: (a) by S-nitrosylation of cysteine residues in target proteins; transcriptional factors and other proteins, (b) phosphorylation of transcription factors or other proteins, and (c) by regulating the cellular localisation of transcription factors. The post-translational modification of proteins can influence the cellular localisation of transcriptional factors. As a result of the above three NO contributes to the regulation of several pathways. A typical response of NO is to activate soluble guanylyl cyclase by binding to its heme group in cells to catalyse the conversion of GTP to cGMP that in turn activates cGMP-dependent protein kinase and other kinases, which mediate many of the normal physiological functions elicited by NO. Further it can also cause ADP ribosylation of proteins. It interacts with NF-kappa B (NF= nuclear factor) and activate it which acts as a transcription factor in influencing iNOS gene expression. Its importance among the general public is because of the drug Sildenafil citrate, popularly known by the trade name Viagra. The drug stimulates penile erections by increasing vasodilation and blood flow, primarily by enhancing signaling through the nitric oxide pathway in the penis. NO acts through the stimulation of the soluble guanylate cyclase, forming cyclic-GMP which in turn enhances the release of neurotransmitters resulting in smooth muscle relaxation and vasodilation. Cyclic-GMP can activate protein kinase G, which causes reuptake of Ca^{2+} and the opening of calcium-activated potassium channels. The fall in concentration of Ca^{2+} ensures that the myosin light-chain kinase (MLCK) can no longer phosphorylate the myosin molecule, ultimately leading to relaxation of the smooth muscle cell. In the heart cGMP brings about less forceful contractions by stimulating ion pumps that expels calcium ions from the cytosol. This NO induced relaxation of heart muscle relieves the person from angina -the pain caused by the contraction of heart muscle because of blocked coronary arteries.

Few Cis-elements involved in NO facilitating the binding of transcription factors to the promoter site and enhancing the transcription have been identified. For example, EF2 element with the sequence—TTTCCCGC (or slight variations of this sequence); GATA transcriptional regulators normally recognise the consensus sequence WGATAR (W = T or A; R = G or A); AP2 element has 'G/ACCGNC' or similar sequences. However, some other few Cis elements can cause repression.

Other Transduction systems

Details regarding sensory transduction systems i.e. in response to smell (odorants) and tastants (sweet, e.g. glucose; salty e.g. sodium ion; umami e.g. glutamate (used in noodles and other Chinese food items); bitter, e.g. quinine; sour, e.g. hydrogen ions (acidic), hearing and touch are not discussed here. Students may read this in some specialised/ advanced courses. A few points about the above are as follows: Odorants apparently function through 7TM receptors, specific G-proteins and adenylate cyclase and cation channel activation. There are a huge number of olfactory (odorant binding) receptors (proteins) in the nasal epithelium. Each odorant can bind to many receptors to different extent and each receptor can bind to many odorants to different extent. The entire combination of receptors-odorant complexes is transmitted by neurons to a specific site in the brain called olfactory bulb and the smell is thereby perceived. Tastants are perceived by a group of cells which constitute taste buds. These buds are attached to sensory neurons. These neurons transmit electric impulses to the brain in response to stimulation by various tastants. Specific 7 TM receptors for sweet, bitter and umami tastes have been identified and hence their signal transduction is apparently mediated by 7 TM receptors. 7 TM receptors are not involved in case of salty tastants. The taste is detected by the passage of tastants through ion channels existing on the surface of the cells of the tongue. A similar situation with sour taste with the difference being that it is the hydrogen ions that pass through the ion channels. Ion channels are also involved in responding to hearing and touch.

32

Signal Transduction in Plants

General Introduction

Hormones controlling plant growth and development are very much different from that of humans and other mammals. In mammalian systems details are well worked out and to the same extent they are not so in plants. Signal Transduction in Plants is an area to be explored. Initially scientists made attempts to detect the various components of the mammalian signal transduction cascades and were successful in detecting several components in different plant systems, *viz.* membrane bound and soluble receptor kinases; GTPases (both heterotrimeric and single polypeptide); 14-3-3 proteins; different protein kinases and phosphatases; inositol phospholipids; calmodulin and other calcium binding proteins; calmodulin binding proteins; cyclic nucleotides; polypeptide growth factors and the phenomenon of zymogen activation in the production of polypeptide growth factors etc. All the above are supposed to be involved in regulating the various physiological processes and responses to various external stimuli. Considerable information was obtained from Arabidopsis, tobacco, tomato, rice, alfalfa and brassica etc. However, the entire cascades and their influence on metabolic pathways and gene expression could not be worked out as in mammalian systems. Plant hormones seem to work in a different manner and their mechanism of action- as to how the final response is achieved- is yet to be worked out, though we know the various processes influenced by various hormones. In plants, it is presumed that, it is not just one but a balance of all the five principal hormones and some other growth regulators that ultimately decide the final response. Signal transduction pathway of each hormone can interact with the signal transduction pathways of other hormones. to regulate different physiological processes. There is a need to study in detail the transcriptomics (expression level of all mRNA's), proteomics (expression of all proteins -enzymatic and non-enzymatic) and metabolomics (expression of all metabolites) in response to signal molecules like hormones to understand the role of various metabolic pathways in bringing about the final response. Particularly in plant tissue culture technology this is very important to know the various processes like root and shoot induction, embryogenesis etc. in response to various hormonal combinations. A brief overview of signal transduction in plants in relation to different signals is given here. Initially the five principal hormones are discussed followed by other molecules known to play a signaling role. They are considered as endogenous signals. Responses to external signals like light, water temperature etc. are not discussed. However, a little coverage of the photoreceptor phytochrome is made. (Further details students may study in some specialised/advanced courses. To start with the book of Russell Jones *et al.* is recommended). Mostly the information is obtained from the crops, rice, tobacco, Arabidopsis, peas, wheat, mung bean, rapeseed, barley, maize, tomato and brinjal.

In general hormones act at transcriptional level by binding to specific Cis-elements in the promoters of the genes distributed on the genome which are referred to as hormone responsive elements or simply response elements. These Cis-elements are short highly conserved sequences which respond to a particular hormone or to other external signals. Many genes possess the same Cis-elements indicating that all these genes respond to a particular hormone or another signal. Alternatively, the same hormone may respond to different Cis-elements. Cis-elements are short sequences which vary in size and composition and in different plants. Other important components and events of signal transduction are phosphorylation and dephosphorylation by specific protein kinases and protein phosphatases and ubiquitin mediated protein degradation of specific proteins.

Role of different principal hormones and other growth regulators and signals in signal transduction

(a) Auxins

Chemically they are indole derivatives and are synthesised from the amino acid tryptophan. Most abundant naturally occurring auxin is IAA (Indole-3-acetic acid). The other important one is IBA (Indole-3-butyric acid). Further there are non-indole auxin like acting metabolites e.g. Phenyl acetic acid. Auxins can exist as conjugates of sugars (glucose) and amino acids (aspartate, glutamate, alanine and leucine). Concentration of the above metabolites vary and accordingly influence the various processes individually and/or in co-ordination with other hormones. Auxins are exploited in agriculture and in tissue culture technology. They regulate many processes; for example, apical dominance, gravitropism, root initiation. In tissue culture a high ratio of auxin to cytokinin promotes rooting.

There are specific Cis-elements in the promoters of auxin responsive genes. A protein termed ARF (auxin responsive factor) binds to these Cis-elements and activate the transcription. When auxin is absent there is no expression of genes, because certain transcriptional repressors called AUX/IAA represses the activity of ARF's. In the presence of auxin, a specific receptor, TIR1 forms a auxin-receptor complex which in turn binds to transcriptional repressors AUX/IAA and subject them to ubiquitin mediated protein degradation by the 26 proteasome (see ubiquitin mediated protein degradation) complex. This process requires the help of some additional protein factors. When the repressors are degraded there is no hindrance for the induction of auxin responsive genes. Other than TIR1 some more auxin binding proteins (ABP's) have been found. By bioinformatic analysis as many as 9 Cis-elements have been identified involved in auxin mediated transcription.

(b) Gibberellins (GA's)

Chemically they are tetracyclic diterpenes having 19 or 20 carbon atoms. Compared to auxin and some other hormones their number is very high. As of 2013, there seems to be chemically 136 Gibberellins but not all are biologically active. They are numbered in order of their discovery. There is a considerable confusion regarding the active forms of Gibberellins present in plants. They are GA1 and GA4. However, it is common to use GA3 originally characterised in fungi in experiments with plants. Green revolution because of dwarf genotypes is because of mutations in gibberellin biosynthesis. Other major responses of Gibberellins are, secretion of hydrolytic enzymes by aleurone cells in cereals, breaking of dormancy and induction of flowering. One classical example is GA induction of alpha-amylase in the aleurone layer. ABA is antagonistic to GA and reverses the responses induced by GA. Cis-elements for GA have been found.

GA-response elements (GARE) in genes coding for hydrolytic enzymes in cereals are composed of TAACAGA, and TAACGTA. Further, two copies of CAACTC also are found. This sequence is termed as CAACTC regulatory elements (CAREs). The same gene may have both CARE and GARE elements. Any mutations in these elements leads to loss of inducibility of genes by GA.

Bioactive GA's bind to GA receptor GID1. This complex recognises specific repressor proteins called DELLA domain proteins. The entire complex is now can be degraded by ubiquitin mediated protein degradation by 26 proteasomes. When once the repressors-DELLA domain proteins- are degraded, GA responsive genes continue to be transcribed. However there seems to be some exceptions where DELLA proteins can increase the transcription of some genes. The concentration of bioactive GA's seem to be regulated by the activities of specific oxidase enzymes. Hydroxylation at carbon 20 and carbon 3 of GA 12, results in the production of bioactive GA's, GA4 and GA1. High concentration of bioactive GA's -GA4 and GA 1 feedback inhibits GA4 oxidase and GA3 oxidase respectively.

(c) Cytokinins (CK's)

They are chemically diverse. Though predominantly they are purine adenine derivatives there do exist aromatic derivatives. Two common examples are diphenyl urea and thidiazuron. Phenyl derivatives are not found in plants. Among synthetic cytokinins also adenine derivatives are the most common. About 200 naturally-occurring or synthetic cytokinins are known, to date. Examples for naturally occurring cytokinins are, iPA (isopentenyl adenine) cis and trans zeatin and dihydrozeatin. Examples for synthetic cytokinins are, kinetin (6-furfurylamino purine), 6-benzyladenine. Biosynthesis of naturally occurring cytokinins occur in plastids.

Major processes regulated by cytokinins are cell division either alone or in conjunction with auxins, shoot and root organogenesis, seed germination, senescence and lateral bud growth.

Protein receptors for cytokinin have been identified. They are primarily found in plasma membrane and can also be found in the membranes of endoplasmic reticulum. The receptors seem to function as dimers. Cytokinin signaling involves phosphorylations. When the hormone binds the receptor, the receptor transfers a phosphate group to a cytoplasmically located protein called AHP (Arabidopsis histidine phospho transfer protein). Of course the prefix Arabidopsis may not be used by others who detect such proteins in cytosol. So the term "cytosol located histidine phospho transfer protein (CHP) is more apt). AHP's in turn transfer their phosphate group and activate another set of proteins called ARR (Arabidopsis response regulatory proteins). These regulatory proteins are transcriptional factors which induces transcription in response to cytokinins. There are two types of response regulatory proteins -Type A and B. Type A when phosphorylated induces the transcription while Type B inhibits the transcription of genes induced by Type A. Parallelly Type B inducing transcription of genes under its control. In addition to the above mentioned transcription factors, there also exist another set of proteins -(CRF's) cytokinin response factors which can respond to cytokinin signaling. Two specific Cis-elements in the promoters of cytokinin-induced genes involved in activating their transcription have been identified using bioinformatic tools- a pentanucleotide motif with the sequence AGAT[T,C] (called CRM - cytokinin response motif) and an octameric sequence AAGAT[T,C]TT (called ECRM-extended cytokinin response motif). In 36 of the 65 promoters of cytokinin induced genes, ECRM is absent.

The concentration of hormone is controlled by (a) reversible conjugation with amino acids and sugars. Conjugation makes the hormone inactive and these are stored in vacuole. Conjugation by glucose takes place at the positions N-7 or N-9 of the purine adenine or on the hydroxyl group of isopentenyl side chain. Amino acid conjugation with alanine can take place at the position N-9 of adenine. Another level of control is to regulate the mobilisation of conjugated cytokinins from vacuole into cytosol. (b) catabolisation of the hormone. The hormone can be broken down to free base adenine and 3-methyl - 2-butenal. This reaction is catalysed by a FAD dependent cytokinin oxidase. In the process FAD is reduced to $FADH_2$. Hence the activity of this enzyme is another regulatory point for maintaining the levels of cytokinin. Higher the activity of the enzyme, lower the levels of cytokinin and vice versa. Any reduction in the hormonal levels, the plant suffers from developmental defects.

(d) Ethylene (C_2H_4)

It basically differs from the other hormones in that ethylene is a gaseous hormone. It is synthesised from methionine and all the enzymes involved in the synthesis are important for manipulating the levels of ethylene. Steps are:

(a) methionine + ATP $\rightarrow$ S-adenosylmethionine + Pi + PPi; This reaction catalysed by SAM synthase.

(b) S-adenosylmethionine $\leftrightarrow$ 1-aminocyclopropane-1-carboxylate (ACC) + methylthioadenosine; This second reaction is catalysed by ACC synthase.

(c) ACC + 1/2 $O_2 \rightarrow CO_2$ + HCN + H_2O + ethylene; This last step is catalysed by ACC oxidase.

Unlike with other hormones catabolism does not seem to be playing a role in regulating the levels of ethylene. Because as it is a gas, it will just diffuse out from the site where it is not required any more.

There are several isoforms of ACS and ACO. Each isoform may be specific or respond differently to a stimulus. ACS gene expression is upregulated by several stimuli. Some examples are; fruit ripening, abiotic stresses particularly water logging conditions, wounding. Further its biosynthesis can in turn be regulated by other hormones auxin and ABA. The rate of degradation of ACS protein is regulated by the phosphorylation status. Phosphorylation at C-terminal end of ACS protein makes it less susceptible to ubiquitin mediated degradation. Thereby higher activity of ACS can be maintained for a longer time.

Ethylene induces ripening in climacteric fruits and in these fruits expression of ACO genes increases during ripening.

The classical response of ethylene is the triple response observed in pea and mung bean i.e. bending and thickening of shoots and pronounced curling of the apical hook. Other major processes influenced by ethylene are; leaf abscission, cell expansion and seed germination (in addition to the above mentioned).

Receptors for ethylene have been identified and all of them have a N-terminal domain that binds to ethylene and a C-terminal end which possesses a histidine protein kinase domain. In the absence of ethylene the ethylene receptors inhibit further ethylene signal transduction events and inhibit the ethylene responses. Ethylene receptors are localised in endoplasmic reticulum. An important receptor is ETR 1. In the presence of ethylene, the receptors are inactivated and can no longer inhibit ethylene responses. In brief the signal transduction events are as follows: In the absence of ethylene ETR 1 leads to a series of protein kinases to operate in succession. These protein kinases are called MAP kinases (MAP= mitogen activated protein). They phosphorylate serine and threonine residues on a protein and hence classified as serine/threonine protein kinases. MAP kinase is activated by phosphorylation by another kinase called MAPKK which in turn is activated by phosphorylation yet by another kinase called MAPKKK. Several target proteins are phosphorylated by MAP kinase cascade. These phosphorylations inactivate the ethylene responses. In the presence of ethylene, the receptor is inactivated and the above successive phosphorylations do not occur. As a result, a membrane localised protein, EIN2 becomes active. The active EIN 2 activates transcription factors which induce transcription of ethylene responsive genes.

Ethylene responsive elements (EREs) have been identified in some crops. They seem to lie in the promoter region spanning "355/"348 and "683/"676. The later was found to be more effective in brinjal. The 8 nucleotide ERE sequence is, ATTTCAAA. Another closely matching sequence AAATTCAAA was observed in tomato. A putative ERE element AATTCAAG was found at "355/"348. In addition to the above sequences, GCC box in promoters also mediate ethylene response influencing other physiological processes. Nuclear proteins interact with these EREs.

(e) ABA (Abscisic acid)

Chemically it is a C15 compound derived from the cleavage of C40 tetraterpenoids in the isoprenoid pathway. Its biosynthesis is regulated by compartmentalisation. Biosynthesis starts with the initial precursor isopentenyl diphosphate in the plastid resulting in xanthoxin. The final two steps from xanthoxin take place in the cytosol *viz.*

(a) Xanthoxin + $NAD^+ \rightarrow NADH + H^+$ + ABA aldehyde. This reaction is catalysed by xanthoxin dehydrogenase also referred to as short chain dehydrogenase.

(b) ABA aldehyde + $H_2O + O_2 \rightarrow$ ABA + H_2O_2. This reaction is catalysed by ABA oxidase or Abscisic aldehyde dehydrogenase.

The levels of hormone are regulated by the relative rates of its synthesis and degradation. It is irreversibly inactivated by getting to degraded to the products, 8'-hydroxy ABA, phaseic acid and dihydrophaseic acid. ABA is also inactivated by conjugation with glucose. But this inactivation is reversible as the glucose ester can be hydrolysed and release free ABA. In the conjugated form it is stored in the vacuole and mobilised into cytosol when needed.

Major physiological responses mediated by ABA are; causing rapid closure of stomata and thereby helping the plant in water stress conditions, synthesis of late embryogenic abundant proteins (LEA) in the seeds during development, causes seed and bud dormancy, and promoting senescence of leaves. ABA can induce ethylene expression which in turn can cause abscission. So ABA responsive genes are diverse and numerous like other hormones.

ABA receptors are proteins with START domains. (START refers to star related lipid transfer and has a domain of about 210 residues which bind to various lipids. The domain is usually at the C-terminal end). These type of receptors are primarily involved in mediating the hormone responses. ABA responses require the functioning of specific protein kinases called SNRK kinases (they are serine/threonine kinases).

These kinases are active only in the presence of ABA and phosphorylation of specific target proteins by them is essential for ABA signal transduction. Important target proteins phosphorylated are, ion channels and transcription factors. Transcription factors bind to ABA response Cis-elements (ABRE) in the promoters of ABA responsive genes and induce transcription. ABRE cis- elements have the conserved sequence—AGCT- and ABRE cis-elements are found in more than 2000 genes. They are termed as, G-box. Though the ACGT is the core sequence conserved in ABREs, the flanking sequences preceding and following ACGT-core may vary. Few examples are; rice, maize and Arabidopsis ABRE has the sequences CGTACGTGTC; GACGTG and CCACGTGG respectively. G-box proteins are predominantly found in ABA induced genes followed by drought and cold and salinity genes. Therefore, it is apparent that ABREs cooperate with other Cis-elements to regulate gene expression under the above mentioned abiotic stresses. For the ABREs to function there is a requirement of additional sequences called coupling elements (CE). CEs are less conserved.

(f) Brassino steroids (BR's)

Chemically they are tetracyclic triterpene molecules also referred to as polyhydroxy steroids, having 27 to 29 carbons. About 70 BR's have been identified and they presumably occur in all tissues. They are required for normal growth and development of the plants. A defect in BR's and their signaling causes severe dwarfing of plants. BR's are involved in promoting cell expansion and cell elongation in conjunction with auxin, promotes vascular differentiation, pollen elongation and pollen tube formation. They also play a role in accelerating senescence and makes the plants to adapt to chilling and drought stress.

The most active form of BR is Brassinolide (BL) which has 28 carbons.

BR's functional activity is dependent on its chemical modifications. Hydroxylation at carbon 2 and 3 of ring A and 22 and 23 of side chain and of the 7 membered B ring is essential for their activity. Hydroxylation is an important step in regulating the levels of bioactive BR's. When BL levels are high it inhibits hydroxylation enzymes which are coded by cytochrome P450. Other modifications like epimerisation, oxidation, sulfonation, conjugation to sugars and amino acids reduce BL activity.

BR's signal transduction involves phosphorylation and dephosphorylation of transcription factors by protein kinase and protein phosphatases respectively. BIN 2 a protein kinase active in its phosphorylated form, phosphorylates several transcription factors involved in inducing BR responsive genes. The transcription factors in their phosphorylated from migrate to cytosol and becomes susceptible to ubiquitin mediated protein degradation and thus no more active. In the presence of BL or other BR's they form a complex with their corresponding receptor BRI 1. The receptor is located in plasma membrane and is a serine/threonine protein kinase. A series of phosphorylations ultimately results in the phosphorylation of protein phosphatase which becomes activated. The activated protein phosphatase dephopshorylates the above mentioned protein kinase BIN 2. When BIN 2 becomes inactive it cannot phosphorylate the transcription factors. Such transcription factors remain in the nucleus and are not destined for ubiquitin mediated protein degradation. So the transcription factors are active and stimulate transcription of BR responsive genes. However, transcription of some genes are repressed. So like other signal transductions -not only in plants but also in mammals-three major sites are involved in BR signal transduction too. Plasma membrane, cytosol and nucleus which are involved in signal binding, signal transduction and regulation of transcription of genes respectively.

Two important transcription factors involved in BR responses are BRZ1 and BES1. BZR1 binds to the Cis-element sequence -CGTG (T/C)G- and is called BRRE (BR Response Element) and is mainly involved in repressing gene expression. Whereas BES1 binds to E-box which has the sequence-CANNTG—which primarily promotes gene expression. Proteins of the category 14-3-3 are implicated in the nuclear import of the above transcription factors.

BR antagonistically with ABA and synergistically with auxin is involved in regulating many physiological processes.

(g) Jasmonates (JA's)

This is the main component responsible for the fragrance of jasmine oil and therefore is used in perfume industry. Jasmonic acid is a 11 carbon compound with a five carbon ring and two side chains.

Jasmonic acid is synthesized from alpha-linolenic acid. The biosynthetic pathway of JA is regulated by compartmentalisation. Three reactions from linolenic acid take place in the plastid and two reactions and three cycles of beta-oxidation take place in peroxisome to give jasmonic acid. Subsequently its conjugation with the amino acid isoleucine (JA-Ile) occurs in cytosol. One more commonly occurring form is methyl jasmonate (MeJA), a volatile compound acting as a signal compound. The addition of methyl group is carried out by the enzyme, SAM-dependent carboxymethyl transferase. As mentioned earlier MeJA can regulate photosynthesis. Methyl jasmonate (see signal transduction) can cause alternative processing of the large subunit of Rubisco. Alternative processing leads to the formation of transcripts with a longer (94 bases) 5′ untranslated region (UTR) and a shorter transcript which is considered as normal form having 59 bases UTR. The longer form inhibits translation. Methyl jasmonate may also destabilise some transcripts of Rubisco small sub-unit.The level of jasmonates is determined by the relative rates of its synthesis and degradation. JA induces many enzymes in its biosynthetic pathway. Its catabolism results in several compounds some of which are biologically active. One important catabolite is tuberonic acid which promotes tuber formation. One of the major physiological responses regulated by jasmonates are; inhibition of seedling growth, inhibition of seed and pollen germination, stimulation of tuber formation and accumulation of storage proteins in seeds, and defense response against herbivory. Their role in seed protein accumulation assumes lot of importance as it can determine grain yield and/or protein content. As they are involved in pollen germination, a deficiency of jasmonates causes the plants to be male sterile. This condition can be reversed by JA treatment.

JA acts by regulating the transcription of a large number of genes and mediates the above physiological responses mentioned above. JAZ proteins are a family of transcriptional repressors. JAZ proteins repress transcription by binding to transcription factors MYC2 and also MYC 3 and 4, required for the expression of JA responsive genes. When JA-Ile binds the SCF-COI1 complex -a multiprotein complex-it facilitates the binding of JAZ to the complex which ultimately results in ubiquitin mediated protein degradation by proteasome. JA and other bioactive derivatives can also bind to the SCF complex. (JA actually binds to the

F-box protein of the above mentioned SCF complex. F-box proteins have a domain of 50 amino acids that mediates protein-protein interaction. So F-box protein may be termed as the JA receptor. The abbreviations of the above mentioned have their own history). When JAZ proteins are degraded there is no more repression of MYC transcription factors. Now transcription of JA responsive genes can occur. High levels of JA-Ile have a feedback inhibition as it can increase the transcription of JAZ genes. G-box (CACGTG) or G-box like motif (core ACGT) has been found in the promoters of many JA-responsive genes. JA-responsive is element (JARE), G(C)TCCTGA, have been identified in some jasmonate responsive genes in Arabidopsis and field mustard. In addition to the above, other JA-responsive Cis-elements are also found: (a) GCC motif which also plays a role in conferring ethylene responsive expression. (b) TGACG sequences in the promoters of some tobacco genes, (iii) CGTCAATGAA and CATACGTCGTCAA sequences in Arabidopsis, (c) TGACG sequences in the promoters of tobacco and barley.

As mentioned under eukaryotic gene regulation, Histone acetyltransferases (HATs) and histone deacetylases (HDACs) regulate gene expression through histone acetylation and deacetylation. HATs and HDACs may interact with other protein complexes which may act either as co-activators or co-repressors. If the above enzymes are not expressed or over expressed, transcription of some JA-responsive genes are altered in Arabidopsis. There are reports that transcription levels of some JA-responsive genes are altered in Arabidopsis.

(h) Salicylic acid (SA)

Chemically it is 2-hydroxy benzenecarboxylic acid. However, it exists in other forms too in plants *viz.* as methyl ester (MeSA), as glucose conjugates, 2-O-beta glucoside (SAG) and beta-glucosyl ester (SGE). It has two biosynthetic pathways which are differentially stimulated. When exposed to abiotic stresses SA is synthesised from chorismate while it is synthesised from trans-cinnamic acid (tCA) in response to pathogen attack. Major physiological responses of SA are induction of flowering, heat generation (thermogenesis) and disease resistance, both localised and systemic and delaying senescence.

SA is an endogenous flowering hormone in Duck weeds. SA induces considerable heat production in Arum spadix. This is possible because of uncoupling of mitochondrial oxidative phosphorylation. Temperature inside the spadix may increase up to 10 degrees centigrade. The advantage to the plant is the high temperature causes the volatilisation of amines. These possess a pungent odour which helps in the attraction of pollinators by the flower. SA can inhibit the ethylene biosynthesis enzyme ACO lowering the ethylene levels. Lowered ethylene levels contribute to a delay in senescence. SA induces defense proteins called pathogenesis related proteins (PR) and contributes to disease resistance. Disease resistance is not only confined to the site of infection but even the parts of the plant distant from the site of infection also develops disease resistance as SA can migrate. This is referred to as systemic disease resistance or systemic acquired resistance (SAR). Hence SA signaling is involved here.

SA induction requires two Cis-elements -SA-responsive Cis-acting elements (TGACG) and W-box—((T)TGAC (C/T)). These elements are found in many SA responsive genes.

(i) Nitric oxide (NO)

(See earlier discussion on this molecule under signal transduction in mammals) It is a gaseous molecule and a free radical. The process of its formation is very much different from that in mammals. It is formed in plants during the process of nitrate assimilation into ammonia as an intermediate reaction:

$NO_2 + 2e^- + 2H^+ \rightarrow NO + H_2O$. (Contrastingly in bacteria NO is formed during denitrification of ammonia (NH_4^+) into nitrogen (N_2).

It is also formed as a result of reduction of nitrous acid HNO_2 by phenolic components and ascorbate in the cell wall;

$HNO_2 + 2R \rightarrow HNO + H_2O + 2R$

The synthesis of NO increases by the application of hormones IAA and ABA. Further hydrogen peroxide (H_2O_2) can stimulate the production of NO in stomatal guard cells. In turn NO can stimulate production of H_2O_2. In fact, stimulation of H_2O_2 by NO is the major mechanism of defense against pathogen attack and thus NO plays a role in disease resistance. Other major processes in which NO plays a role are; stomatal closure, flowering, seed germination and response to abiotic stresses. NO is implicated in responses mediated by SA, JA and ethylene.

NO may cause S-nitrosylation of cysteine in transcription factors and regulate transcription. Several Cis-elements have been found in Arabidopsis for NO regulated genes. Cis-elements may either increase or decrease transcription.

(j) Phytochrome

It is a red light photoreceptor involved in the regulation of several physiological processes. Chemically phytochrome is a biliprotein, consisting of a 120 kDa apoprotein and a linear chain tetrapyrrole which is attached by its A ring to the apoprotein cysteine residue in the apoprotein. The tetrapyrrole which acts as a chromophore is referred to as phytochromobilin. The chromophore can exist in two molecular configurations; a red absorbing form PR (which absorbs light around 600 nm) and a far red absorbing form PFR (which absorbs light around 730 nm).They are referred to as Cis and trans isomers respectively and involves a change in the double bond between carbon 15 and 16 in the D ring (fourth ring). PR and PFR forms are interconvertible. Red light converts PR into PFR and far red light converts PFR to PR. Even in the absence of far red light PFR can be slowly converted to PR under dark conditions. When the forms are reversed the physiological process also is reversed. Active phytochrome exists as a dimer.

Phytochrome regulates several plant processes. A few important examples are; induction of germination of certain seeds, increasing chlorophyll synthesis and plays an important role in flowering, photoperiodism and leaf growth.

When PR is converted into PFR, PFR moves from cytosol into nucleus. In the nucleus PFR interacts with specific proteins called phytochrome interacting proteins (PIF's). There seems to be more than 25 genes coding for phytochrome interacting factors. After the formation of PFR-PIF complex, PIF3 is phosphorylated and is subjected to ubiquitin mediated protein degradation. This degradation favours the transcription of genes involved in photomorphogenesis i.e. response of plants to information in light. Many Cis-regulatory elements -also called light responsive elements (LRE) -have been identified in different crops. Some of these elements are involved in phytochrome mediated regulation. Few examples are GT elements with the sequence—GGTAA, GGTTAA GGTAATT. In addition to GT they are rich in A and the length of the sequence can vary in size and composition. G-box elements and GATA elements are also involved and have been discussed earlier. A single transcription factor can bind to many related GT elements. Some Cis elements are involved in enhancing the transcription while some are involved in decreasing the transcription. A specific 12 bp Cis-element sequence—GGATTTTACAGT—has been identified in pea (*Pisum sativum*) which is involved in the down regulation of a GTPase gene mediated by phytochrome.

(k) Photoreceptors also exist for blue/UV light and plants also respond to green light

These details students may read in specialised/advanced courses. Suffice to mention here that all blue light photoreceptors are proteins containing the chrompophores FAD and pterin and the signal transduction involves protein phosphorylation and ubiquitin mediated protein degradation. UV light is damaging to nucleic acids and proteins and specific UV sensitive photoreceptors have been identified.

(l) Strigolactones and polyamines

Suffice to mention here that strigolactones are chemically terpenoid lactones derived from carotenoids and have four rings. They are found in root exudate signals and can trigger parasitic weed seed germination, and also inhibits shoot branching. Strigolactones in the root exudates stimulate the branching and growth of symbiotic arbuscular vesicular mycorrhizal fungi promoting plant-fungus symbiosis. When plant is suffering from phosphate deficiency this symbiosis will help the plant to gain phosphate with the help of symbiotic fungus. Similarly, different flavonoids exuded from the legume roots induce nodulation genes in rhizobium which fixes atmospheric nitrogen into ammonia, and promotes symbiotic nodule formation in legumes. However, some flavonoids acts as anti-inducers. Polyamines are small metabolites which have two or more amino groups. They are synthesised from ornithine and arginine and lysine. As compared to other principal hormones they are found in the cells at much higher levels. So if one is particular about attributing "existence in a low concentration" as a criteria for the definition of an hormone as in case of the five principal hormones, then they cannot be classified so. Three major polyamines are putrescine, spermine and spermidine. Two other important examples are cadaverine and thiospermine. Tissue culture technologists use polyamines as they play an important role in the regulation of somatic embryogenesis. Being positively charged they bind to DNA.

SAM (S-adenosylmethionine) is used both in the biosynthetic pathways of polyamines and ethylene biosyntheis. Hence these two biosynthetic pathways are competitive inhibitors to each other and may regulate the level of both ethylene and polyamines and consequently their responses on plant processes.

(m) A brief description of a few polypeptide growth factors unique to plants is described here (Students should be clear that the following information has not entered fully into all the text books and the information has been compiled from published papers).

(a) Systemin

It is a short polypeptide of 18 amino acids resulting from the proteolytic cleavage of a larger precursor of 200 amino acids called Pro-systemin. Though it is a short polypeptide the sequence appears to be varying among the plants. In general, the N-terminal 14 amino acids binds to the receptor while the last four -terminal amino acids actually play a role in signal transduction. Systemin plays a role in defense responses against insect attack particularly by inducing proteinase inhibitors. Specific alterations in the amino acid sequence and antisense inhibition of Pro-systemin can affect the above response.

(b) RALF (rapid alkalisation factor)

It is a 49 amino acids polypeptide formed from its precursor of 115 amino acids. Polypeptide and its binding proteins are widely distributed among plants. It is involved in the inhibition of cell division and root growth.

(c) Phytosulfokines

They are very small oligopeptides of 4 or 5 amino acids with their tyrosine residues sulfated. Sulfation is essential for their activity in regulating cell proliferation and cell differentiation.

(d) CLAVATA polypeptides

They are extracellular signaling peptides of 79 amino acids involved in the regulation of cell differentiation.

(e) S-locus signaling polypeptides

Their size varies from 47 to 60 amino acids and are cysteine rich. They interact with serine/Threonine receptor protein kinases and are implicated in pollen self-incompatibility in Brassicaceae family.

In addition to the above several short peptides have been identified in various plants which are likely to play an important role in plant growth, development and differentiation. Confirming the presence and role of the above polypeptide growth factors is an important area of research to explore.

References

I. General

Arumugam, S., Rao, A. S. and Rao, M.V. (2003). *In vitro* propagation *of Aegle marmelos(*L.*)Corr., a medicinal tree.* In: Micropropagation of woody trees and fruits. (Eds). Jain, S. M. and Ishii, K. Kluwer Academic Publishers, Dordrecht, The Netherlands. pp. 269-315.

Bakrudeen Ali Ahmed, A., Rao, A. S. and Rao, M.V. (2008). Role of *in vivo* leaf and *in vitro* callus of *Gymnema sylvestre* in maintaining the normal levels of blood glucose and lipid profile in diabetic Wistar rats. Biomedicine. 28: 134-138.

Bakrudeen Ali Ahmed, A., Rao, A. S. and Rao, M.V. (2010). *in vitro* callus and *in vivo* leaf extract of *Gymnema sylvestre* stimulate beta-cells regeneration and antidiabetic activity in Wistar rats. Phytomedicine. 17: 1033-1039.

Barrett, A.J., N.D. Rawlings, N. D. and Woessner, J. F. (1998). Handbook of Proteolytic Enzymes. Academic Press, New York.

Benjamin, L, Schleiff , W. E. and Zerges, W. (2013). Protein targeting to subcellular organelles via mRNA localisation. Biochimica et Biophysica Acta—Molecular Cell Research- 1833: 260-273.

Berg, J. M, Tymoczko, J.L. and L. Stryer, L. (2012). Biochemistry (7th Edition), W.H. Freeman & Palgrave Macmillan, New York, USA (students may also find useful the 6th edition also of 2006).

Berendzen, K. W., Weiste, C., Wanke, D., Kilian, J., Harter, K. and Dröge-Laser, W. (2012). Bioinformatic *cis*-element analyses performed in *Arabidopsis* and rice disclose bZIP- and MYB-related binding sites as potential AuxRE-coupling elements in auxin-mediated transcription. BMC Plant Biology 12: 125-143.

Bonner, E.R., Cahoon, R. E., Knapke, S. M. and Jez, T.M. (2005). Molecular basis of cysteine biosynthesis in plants—Structural and functional analysis of O-aetylserine sulfhydrylase from *Arabidopsis thaliana*—J. Biol.Chem. 280: 38803-38813.

Brenner, W. G. and Schmülling, T. (2015). Summarizing and exploring data of a decade of cytokinin-related transcriptomics. Front Plant Sci.6: 29-41.

Bromke, M. A. (2013). Amino Acid Biosynthesis Pathways in Diatoms. Metabolites, 3: 294–311.

Brown, T. A. (2002). Genomes, 2nd Edition, Oxford: Wiley-Liss, UK.

Brown, T. A. (2006). Genomes, 3rd edition, Garland Science Taylor & Francis Group, UK/USA.

Buchnan, B. Gruissem, B. W. and Jones, R.L. (2000). Biochemistry and molecular biology of plants. American society of plant biologists, USA. (second edition of this book-2015).

Campbell, M. K. and Farrell, S. O. (2006). Biochemistry 5th Edition, Cenaga Learning, Australia.

Campbell, M. K. and Farrell, S. O. (2012). Biochemistry. Brooks Cole, Australia.

Chastain, C. J., Failing, C. J., Manandhar, L., Zimmerman, M. A. Lakner, M. M. and Nguyen, T.H.T. (2011). Functional Evolution of C4 Pyruvate, Orthophosphate Dikinase. J. Exp. Bot. 62: 3083–3091.

Chetal, S., Wagle, D.S. and Nainawatee, H. S. (1982). Lipid changes in two cultivars of wheat and barley during leaf development. Z. Pflanzenphysiol. 107: 187-191.

Chetal, S., Wagle, D.S. and Nainawatee, H. S. (1983). Phospholipid changes in wheat and barley chloroplasts under water stress. Plant Science Letters, 29: 273-278.

Cho, M.H, Jang, A., Bhoo, S. H., Jeon, J. S, and Hahn, T. R. (2012) Manipulation of triose phosphate/phosphate translocator and cytosolic fructose-1,6-bisphosphatase, the key components in photosynthetic sucrose synthesis, enhances the source capacity of transgenic Arabidopsis plants. Photosynth Res. 111: 261-268.

Clark, G. B., Thompson, Jr. and Roux, S. J. (2001). Signal trans-duction mechanisms in plants:An overview. Current Science (Bengalur-India), 80: 170-177. (This issue, number 2, 25 January, has a special section on plant molecular biology and plant signal transduction with 14 articles, pp. 143-279).

Cox, M. M. and Nelson, D. L. (2008). Lehninger-Priciples of Biochemistry, 5th Edition, W.H. Freeman and company, New York, USA.

Garrett, R. H. Grisham, C. M. (1995). Biochemistry, Saunders College Pub: USA.

Glazer, A. N. and Nikaido, H. (1995 and 2007), Microbial Biotechnology: Fundamentals of Applied Microbiology. Cambridge University Press, UK.

Hartl, D. L. and Jones, E. W. (2009). Genetics—Analysis of genes and genomes. Jones and Bartlett Publishers, Inc. USA.

Dao-Xiu Zhou (1999). Regulatory mechanism of plant gene transcription by GT-elements and GT-factors Trend Plant Sci. 4: 210–214.

de la Torre, F., Rafael, A., Cañas, M., Pascual, B., Avila, C. and Cánovas, F. M. (2014). Plastidic aspartate aminotransferases and the biosynthesis of essential amino acids in plants. Journal of Experimental Botany doi:10.1093/jxb/eru240.

Dennis, D. T. and Turpin, D. H. (1990). Plant physiology, biochemistry and molecular biology. Longman, U.K.

Dennis, D. T., Turpin, D. H. and Lefebvre, D. D and Layzell, D. B. (Editors)(1997). Plant metabolism, 2nd Edition, Longman, U.K.

Devlin, T. M. (Eds.) (2001). Textbook of Biochemistry: With Clinical Correlations (5th Ed.); John Wiley & Sons, Inc., New York, USA.

Dey, P. M. and Harborne, J. B. (Eds) (1997). Plant Biochemistry. Academic press, USA.

Dhakshinamoorthy, S, Sridharan, S.R., Li, L., Ng, P.Y., Boxer, M.L. and Porter, A. G. (2007). Protein/DNA arrays identify nitric oxide-regulated *cis*-element and *trans*-factor activities some of which govern neuroblastoma cell viability. Nucleic Acids Res. 35: 5439–5451.

Feilleux, -Dneha, S., Garletti, M., Aggarbeck, M. Polyard, M., Bouquet, J., Hanorme, J. and Barouki, R. (1993). Ell specific regulation of aspartate aminotransferase by gluoortioids in the rat kidney. Amer. J. Physiol. 265(5Pt1): C1 298–305.

Hames, D. and Hooper, N. (2011). BIOS Instant Notes - Biochemistry 4th Edition, Taylor & Francis Ltd, U.K.

Heldt, H.W. and Piechulla, B. (2011), Plant Biochemistry-Academic Press USA (Translation from 4th German edition 2008).

Hess, D. T., Matsumoto, A, Kim, S. O., Marshall, H. E, and Stamler, J. S. (2005). Protein S-nitrosylation: purview and parameters. Nat Rev Mol Cell Biol. 6: 150-66.

Iglesias, A. A., Gonzalez, D. and Andreo, C. (2010). The pathway of photosynthesis and its regulation. Biochemical education, 14: 98-102.

Iglesias, A. A., Gonzalez, D.H. and Andreo, C.S. (2010). Crassulacean acid metabolism: A pathway for synthetic CO_2 fixation in arid habitats. Biochemical education. 15: 111-115.

Inaba, T., Nagano,Y., Sakakibara, T. and Sasaki, Y (1999). Identification of a *cis*-Regulatory Element Involved in Phytochrome Down-Regulated Expression of the Pea Small GTPase Gene *pra* 2. Plant Physiol. 120: 491-500.

Jain, S., Nainawatee, H. S., Jain, R.K. and Chowdhury, J. B. (1991). Proline status of genetically stable salt tolerant *Brassica Juncea* L. somaclones and their parent cv. Prakash. Plant & Cell Report, 9: 684-687.

Jain, V., Sheoran, I.S. and Nainawatee, H. S. (2002). Effect of rhizospheric application of naringenin and nitrate on enzymes of ammonia assimilation and carbohydrate metabolism in pea (*Pisum sativum* L.) – *Rhizobium leguminosarum* biovar *viciae* symbiosis. J. Plant Biol. 29: 265-270.

Jeong, M. J., and Shih, M. C. (2003) Interaction of a GATA factor with cis-acting elements involved in light regulation of nuclear genes encoding chloroplast glyceraldehyde-3-phosphate dehydrogenase in Arabidopsis. Biochem Biophys Res Commun. 300: 555-562.

Jones, R., Oughama, H., Thomas, H. and Waaland, S. (2013). The molecular life of plants—American Society of Plant Biologists-Wiley-Blackwwell, USA.

Kendziorek, M., Paszowski, A. and Zagdanska, B.(2012). Differential regulation of alanine amino transferase homologues by abiotic stresses in wheat (*Triticum aestivum L.*) seedlings. Plant cell. Rep. 31: 1105-1117.

Krebs, J. E., Goldstein, E. S. and Kilpatric, S. T. (2014), Lewin's Genes XI Jones & Bartlett Learning, USA (Genes X and IX also would be informative).

Kumar, S., Verma, R.K., Mishra, B.K. and Chopra, D. S. (2014). Role of Enzymes in food processing. In "Food and Nutritional Security by Sustainable Agriculture" pp. 453-477. Eds. Mishra, B.K., Sunil Kumar, S. and Kumar, J. New India Publishing Agency, New Delhi.Web: www.nipabooks.com

Latchman, D. S. (2005/2007). Gene Regulation - A eukaryotic perspective; 5th Edition; Taylor& Francis Group, US/UK).

Lea, U.S. Leydecker, M-T, I. Quilleré, I, Meyer, C and Lillo, C. (2006). Post-translational Regulation of Nitrate Reductase Strongly Affects the Levels of Free Amino Acids and Nitrate, whereas Transcriptional Regulation Has Only Minor Influence. Plant Physiol. 2006; 140: 1085–1094.

Leegood, R. C. and Walker, R. P. (2003). Regulation and roles of phosphoenolpyruvate carboxykinase in plants. Arch Biochem Biophys. 15: 204-210.

Lehner, B, Williams, G, Campbell, RD, Sanderson, CM (2002). Antisense transcripts in the human genome. Trends Genet. 18: 63-65.

Li, H. Y., Wei, W. and Li, Y. (2006). Roles of salicylic acid-responsive cis-acting elements and W-boxes in salicylic acid induction of VCH3 promoter in transgenic tobaccos. Acta Biochim Biophys Sin (Shanghai). 2006, 38: 46-52.

Li, J., Li, G., Wang, H. and Deng, X.W. (2011). Arabidopsis Book. (American society for plant biologists), 9. e0148.

Liu, C, Vitnary Choe, V. and Rao, H (2010). Genome wide approaches to systematically identify substrates of the ubiquitin-proteasome pathway. Trends in Biotechnology-28: 461-467.

Mack, G. S. (2007). MicroRNA gets down to business. (2007). Nat. Biotechnol. 25: 631-638.

Madan, S., Nainawatee, H.S., Jain, R.K. and Chowdhury, J.B. (1995). Proline and proline metabolizing enzymes in *in vitro* selected NaCl - tolerant *Brassica juncea* L. under salt stress. Annals of Botany, 76: 51-57.

Mondal, K., Saxena, N.; Malhotra, Sarla, P. and Singh, R. (2008). Possible interactions of polyamines and ethylene during ripening of guava (psidium guajava l.) Fruits. J. Food Biochemistry 32: 46-59.

Moreno-Sánchez, R, Saavedra, E, Rodríguez-Enríquez, S, Olín-Sandoval, V. (2008). Metabolic control analysis: a tool for designing strategies to manipulate metabolic pathways. J Biomed Biotechnol. 2008; 2008:597913. doi: 10.1155/2008/597913.

Mur, L. A. J. et al (2013). Nitric oxide in plants: an assessment of the current state of knowledge. AoB PLANTS (2013) 5 : pls052doi: 10.1093/aobpla/pls052.

Mur, L.A.J., Prats, E., Pierre, S., Hall, M. A. and Hebelstrup, K.H.(2013). Integrating nitric oxide into salicylic acid and jasmonic acid/ethylene plant defense pathways. Front. Plant Sci., 27 June 2013 | http://dx.doi.org/10.3389/fpls.2013.00215.

Murray, R. K., Granner, D. K. and V. W. Rodwell, V. W. (2006). Harpers's illustrated Biochemistry (27th Edition). McGraw-Hill Education (Asia).

Murray, K. R., David A. Bender, D. A., Botham, K. M., Kennelly, P.J., Rodwell, V. W. and Anthony Weil, P., Harper's Illustrated Biochemistry (2012 and 2015) 29th and 30th Editions—The McGraw-Hill Companies, Inc. USA.

Ochs, R. S. (2014) Biochemistry, Jones & Bartlett, USA.

Onouchi, H., Lambein, I., Sakurai, R., Suzuki, A., Chiba, Y. and Naito S. (2004). Autoregulation of the gene for cystathionine gamma-synthase in Arabidopsis: post-transcriptional regulation induced by S-adenosylmethionine. Biochem Soc Trans. 32(Pt 4): 597-600.

Pagani, C. C., Bauler, M., Narayanan, R., Miller, D.D. and Dalton, J. T. (2012). Alanine amino transferase regulation by androgens in non-hepatic tissues. Pharmaceutical research. 29: 1046-1056.

Palmieri, M. C. et al. (2008) Nitric oxide-responsive genes and promoters in *Arabidopsis thaliana*: a bioinformatics approach. J. Exp. Bot. 59, 177-186.

Primrose, S. B and Twyman, R. M. (2006). Principles of gene manipulation and genomics (7th Edition), Blackwell Publishing, U.K.

Pal. A.P.; Malhotra, Sarla P.; Saharan, M.R. and Singh, R. (2004). Purification and characterisation of hexokinase from developing seeds of Brassica campestris L Physiol Mol. Biol. Plants. 10 : 217-222.

Rao, A. S. (1996). Terminology in microbial metabolism. Biochemical education 24: 61. (Present name of the journal is Biochemistry and Molecular Biology Education (abbreviated as BAMBED).

Rao, S. Adhikarla (2001). Chloroplast genetic engineering. Trend. Biotechnol. 19: 6.

Rao, S. Adhikarla (2017). Some points to ponder on some biotechnologies of current and future relevance. Adv. Biotech. Microbiol.(U.S.A), 3: (issue 3, May 2017), DOI: 10. 19080/ AIBM. 2017.03.555613.

Rao, S. Adhikarla (2016). Conversion of C-3 plants into C-4 plants may lower the Rubisco amounts with negative effects on grain yield and protein concentration. J. Plant. Biochem. Biotechnol. (Springer) DOI10.1007/s 13562-016-0379-9.

Rao, A. S. and Randhir Singh (1983). Theoretical approaches for reducing bioenergetic costs and enhancing plant productivity. Journal of Theoretical Biology, 104: 113-120.

Rao, A. S. and Singh, R. (1986). Improving grain protein quality by genetic engineering—some biochemical considerations. Trend. Biotechnol. 4: 108-110.

Rao, A. S. and Randhir Singh (1988). Ureide metabolism in nodulated legumes. In Advances in frontier areas of plant biochemistry- Prentice -Hall of India, New Delhi, 281-316.

Rawat, R., Xu, Z-F., Yao, K-M. and Chye, M-L. (2005). Identification of *cis*-elements for ethylene and circadian regulation of the *Solanum melongena* gene encoding cysteine proteinase. Plant Mol. Biol. 57: 629-643.

Riesmeier, J. W. (1993). Antisense repression of the chloroplast triose phosphate translocator affects carbon partitioning in transgenic potato plants. Proc Natl Acad Sci USA. 90:6160-6164.

Reyes, J.C., Isabel Muro-Pastor, M. and Florencio, F. J. The GATA Family of Transcription Factors in Arabidopsis and Rice.(2004). Plant Physiol. 134: 1718–1732.

Robert, K., Kulis-Horn, Persicke, M. Jörn Kalinowski, J. (2013). Histi-dine biosynthesis, its regulation and biotechnological application in *Corynebacterium glutamicum* Microbial Biotechnology. 7 : 5–25; Jan PMC3896937(open access journal).

Sabari, B. R., Zhang, Di, David Allis, C. and Zhao, Y.(2016). Metabolic regulation of gene expression through histone acylations. Nature Reviews, Molecular cell Biology, Macmillan Publishers—Online, 7 Dec., 2016. Doi1038/nrm 2016, 140.

Sacco, F, Perfetto, L. and Castagnoli, L. and Cesareni, G. (2012). The human phosphatase interactome: An intricate family portrait. FEBS Letters, 586: 2732-2739.

Sasaki, K., Mitsuhara, I., Seo, S., Ito, H., Matsui, H. and Ohashi, Y. (2007). Two novel AP2/ERF domain proteins interact with *cis*-element VWRE for wound-induced expression of the Tobacco *tpoxN1* gene. The Plant Journal. 50:1079-1092.

Seo, J. S. et al., (2013). Identification of a novel jasmonate-responsive element in the AtJMT promoter and its binding protein for AtJMT repression. PLoS One. 8: e55482. doi: 10.1371/journal.pone.0055482.

Shabnam, Saharan, M.R.; Singh, R.(1993). Purification and characterisation of Phosphoribulokinase from Immature Pods of *Brassica.* Journal of Plant Biochemistry and Biotechnology, 2: 121–124.

Shi, Y. (2009). Serine/Threonine Phosphatases: Mechanism through Structure. Cell, 139: 1-17.

Singh, R. and Malhotra, Sarla P. (2002). Designer crops. In "Recent Advances in Agronomy" (G. Singh, JS Kolar and HS Sekhon Eds). Indian Soc. Agronomy, IARI, New Delhi, pp. 415-445.

Slater, A, Scott, N. W. and Fowler, M. R. (2003 and 2008). Plant Biotechnology—The genetic manipulation of plants. Oxford University Press, UK.

Slocum, D. (2005). Genes, enzymes and regulation of arginine biosynthesis in plants. Plant Physiol. Biochem. 43: 729-745.

Söderberg, M., Raffalli-Mathieu, F. and Lang, M. A. (2007). Identification of a regulatory *cis*-element within the 32-untranslated region of the murine inducible nitric oxide synthase (iNOS) mRNA; interaction with heterogeneous nuclear ribonucleoproteins I and L and role in the iNOS gene expression. Molecular Immunology. 44: 434–442.

Stauffer, G. V. (2004). Regulation of serine, glycine and one-carbon biosynthesis. EcoSalPlus.doi:10.1128/ecosaplus.3.6.1.2. Downloaded from www.asmscience.org

Sutoh, K., and Yamauchi, D. (2003). Two cis-acting elements necessary and sufficient for gibberellin-upregulated proteinase expression in rice seeds. Plant J. 34: 635-645.

Terzaghi, W. B. and Cashmore, A. R. (1995), Light-regulated transcription. Ann. Rev. Pl. Physiol. Plant Mol. Biol. 46: 445-474.

Trewavas, A. (2002). Plant Cell signal transduction: The emerging phenotype. The Plant Cell 14: S3-S4, Supplement, DOI 10. 1105/ tpc.141360(This is a whole special issue on signal transduction, pp. 1-417).

Voet, D. and Voet, J. G. Biochemistry (2004 and 2011), 3rd and 4th Editions, John Wiley & Sons, Inc. USA.

Wadsworth, G.J. (1997). The plant aspartate aminotransferase gene family Physiologia plantarum. 100: 998-1006.

Wang, H. and Deng, X. W. (2004). Phytochrome signaling mechanism -Arabidopsis book (American society for plant biologists) 3: e0074.1.

Wendish, V.F. (2006)(Ed.). Aminoaid biosynthesis—Pathways, Regulation and Metabolic engineering. Springer-Verlag, Berlin.

Winter, G., Todd, G. D. and Funck, D. (2015). Physiological implications of arginine metabolism in plants. Front. Plant sci. 6: 534-547.

Yelin, R. et al (2003). Widespread occurrence of antisense transcription in the human genome. Nat Biotechnol. 21: 379-386.

Zhang, W, Rua, J., Ho, T.H., You, Y., Yu, T. and Quatrano, R. S. (2005). Cis-regulatory element based targeted gene finding: genome-wide identification of abscisic acid- and abiotic stress-responsive genes in Arabidopsis thaliana. Bioinformatics, 21: 3074-3081.

Zhang, X and Liu, C-J. (2015). Multifaceted regulations of gateway enzyme phenylalanine ammonia lyase in the biosynthesis of phenylpropanoids. Molecular Plant; 8: 17-27.

II. e-references

http;//www.biotecharticles.com/Biology-Article/Calvin-Cycle-Regulation-and -Effect-on-Photosynthesis-314.html

http://ejournal.sinica.edu.tw/bbas/content/2005/4/Bot464-01.html

http://www.ncbi.nlm.nih.gov/pmc/articles/PMC4197637/

III. Journal References

1. **Biochemistry and Molecular Biology Education** (abbreviated as BAMBED and formerly Biochemical education). A journal of International Union of Biochemistry and Molecular Biology, Phillip Ortiz (Present editor). Inc. John Wiley & Sons, Inc. USA. (a highly useful journal for students).
2. **Metabolic Engineering**- An official journal of the International Metabolic Engineering Society, Greg Stephanopoulos (Present Editor-in-Chief), Elsevier B. V. Netherlands. (This journal is a must for students to understand metabolic pathways, their regulation and manipulation by various ways to achieve the desired biotechnological goals).

IV. References of Nobel Prizes

1. Andrew, Z. Fire and Craig, C. Mello. The Nobel Prize in Physiology or Medicine 2006 "for their discovery of RNA interference - gene silencing by double-stranded RNA".
2. Richard Axel and Linda B. Buck. The Nobel Prize in Physiology or Medicine 2004 "for their discoveries of odorant receptors and the organisation of the olfactory system".
3. Leland, H. Hartwell, Tim Hunt and Sir Paul M. Nurse. The Nobel Prize in Physiology or Medicine 2001 "for their discoveries of key regulators of the cell cycle".

4. Arvid Carlsson, Paul Greengard and Eric R. Kandel. The Nobel Prize in Physiology or Medicine 2000 "for their discoveries concerning signal transduction in the nervous system".
5. Günter Blobel. The Nobel Prize in Physiology or Medicine 1999. "for the discovery that proteins have intrinsic signals that govern their transport and localisation in the cell".
6. Robert, F. Furchgott, Louis, J. Ignarro and Ferid Murad. The Nobel Prize in Physiology or Medicine 1998 "for their discoveries concerning nitric oxide as a signaling molecule in the cardiovascular system".
7. Alfred, G. Gilman and Martin Rodbell. The Nobel Prize in Physiology or Medicine 1994 "for their discovery of G-proteins and the role of these proteins in signal transduction in cells".
8. Richard, J. Roberts and Phillip, A. Sharp. The Nobel Prize in Physiology or Medicine 1993 "for their discoveries of split genes".
9. Edmond, H. Fischer and Edwin G. Krebs. The Nobel Prize in Physiology or Medicine 1992 "for their discoveries concerning reversible protein phosphorylation as a biological regulatory mechanism".
10. Michael, S. Brown and Joseph, L. Goldstein. The Nobel Prize in Physiology or Medicine 1985 "for their discoveries concerning the regulation of cholesterol metabolism".
11. Earl, W. Sutherland, Jr. The Nobel Prize in Physiology or Medicine 1971 "for his discoveries concerning the mechanisms of the action of hormones".
12. Konrad Bloch and Feodor Lynen. The Nobel Prize in Physiology or Medicine 1964 "for their discoveries concerning the mechanism and regulation of the cholesterol and fatty acid metabolism".
13. Hans Adolf Krebs. The Nobel Prize in Physiology or Medicine 1953 "for his discovery of the citric acid cycle".
14. Carl Ferdinand Cori and Gerty Theresa Cori, née Radnitz. The Nobel Prize in Physiology or Medicine 1947 "for their discovery of the course of the catalytic conversion of glycogen", and Bernardo Alberto Houssay "for his discovery of the part played by the hormone of the anterior pituitary lobe in the metabolism of sugar".

15. Robert, J. Lefkowitz and Brian K. Kobilka. The Nobel Prize in Chemistry 2012 "for studies of G-protein-coupled receptors".
16. Roger, D. Kornberg. The Nobel Prize in Chemistry 2006 "for his studies of the molecular basis of eukaryotic transcription".
17. Aaron Ciechanover, Avram Hershko and Irwin Rose. The Nobel Prize in Chemistry 2004 "for the discovery of ubiquitin-mediated protein degradation".
18. Peter Agre. "The Nobel Prize in Chemistry 2003. "for the discovery of water channels" and Roderick MacKinnon "for structural and mechanistic studies of ion channels".
19. Kary, B. Mullis. The Nobel Prize in Chemistry 1993 "for his invention of the polymerase chain reaction (PCR) method" and Michael Smith "for his fundamental contributions to the establishment of oligonucleotide-based, site-directed mutagenesis and its development for protein studies".
20. Sidney Altman and Thomas R. Cech. The Nobel Prize in Chemistry 1989 "for their discovery of catalytic properties of RNA".
21. Paul Berg. The Nobel Prize in Chemistry 1980 "for his fundamental studies of the biochemistry of nucleic acids, with particular regard to recombinant-DNA" and Walter Gilbert and Frederick Sanger "for their contributions concerning the determination of base sequences in nucleic acids".
22. Luis, F. Leloir. The Nobel Prize in Chemistry 1970 "for his discovery of sugar nucleotides and their role in the biosynthesis of carbohydrates".
23. Melvin Calvin. The Nobel Prize in Chemistry 1961 "for his research on the carbon dioxide assimilation in plants".
24. Arthur Harden and Hans Karl August Simon von Euler-Chelpin. The Nobel Prize in Chemistry 1929 "for their investigations on the fermentation of sugar and fermentative enzymes".

Index

Part - I

Basic Concepts/Types of Metabolic Regulation

Part - II

Carbohydrate metabolism

Gluconeogenesis

Glycogen degradation and Biosynthesis

Part - III
Lipid Metabolism

Fatty acid and Triacylglycerol Biosynthesis

Cholesterol Biosynthesis

Part - IV
Amino acid Metabolism

Glutamate Family (Glutamate, Glutamine, Proline, Arginine)

Aromatic amino acid biosynthesis (Phenylalanine, Tyrosine, Tryptophan)

Aspartate family of amino acids - Aspartate, Asparagine, Methio-nine, Lysine, Threonine

Pyruvate Family of amino acids - Isoleucine, Leucine and Valine, Alanine

Serine family Amino acids- (Serine, Glycine and Cysteine)

Histidine

Strategies for Over Production of Amino acids

Urea Cycle

Part - V
Nucleic acid metabolism
Purine Biosynthesis

Pyrimidine Biosynthesis

Conversion of Ribonucleotides into deoxyribonucleotides

Part - VI
Few Important Metabolic pathways in Plants

Sucrose and Starch Biosynthesis

Nitrate assimilation

Phenylalanine ammonia lyase

PART - VII
Zymogen activation, Protein targeting and Protein turnover

Zymogen activation

Part - VIII
Brief Introduction to Gene Regulation in Prokaryotes and Eukaryotes Gene Regulation in Prokaryotes and Eukaryotes

Signal Transduction in Mammals

Signal Transduction in Plants